Science Says

an exception is really according to order." —JOHANN
WOLFGANG VON GOETHE "We live in a scientific age, yet we assume that
nowledge of science

Science Says

eings, isolated
nd priestlike in their   e are the
materials of life **A Collection of Quotations on the History,** ow and the
why for everythi **Meaning, and Practice of Science**   theorist
s not to be envied. For Nature, or more precisely experiment, is an
nexorable and not very friendly judge of his work. It never says
Yes" to a theory. In the Edited by **Rob Kaplan** says "Maybe," and in
he great majority of cases simply "No." If an experiment agrees with
theory it means for the latter "Maybe," and if it does not agree it
neans "No." Probably every theory will someday experience its "No"—
nost theories, soon after conception. —ALBERT EINSTEIN **"The**
Dark Ages may return on the gleaming wings of Science."
—WINSTON CHURCHILL "Genius is o☐ percent inspiration and
inety-nine percent pe

A Stonesong Press Book   LVA EDISON
W. H. Freeman and Company
Science moves with the spiri New York aventure characterized
oth by youthful arrogance and by the belief that the truth, once

Text design: Diana Blume

Library of Congress Cataloging-in-Publication Data

Science says: a collection of quotations on the history, meaning, and practice of science/edited by Rob Kaplan.
    p. cm.
Includes index.
ISBN 0-7167-4112-1
    1. Science—Quotations, maxims, etc. I. Kaplan, Rob.
Q173.S426 2000
500—dc21

                                      00-042222

Printed in the United States of America

First printing 2000

About the Author

Rob Kaplan has over twenty-five years experience in book publishing. He has held senior-level editorial positions with several major New York–based publishing houses and currently heads his own literary services firm, Rob Kaplan Associates, which he founded in 1998. He is the co-editor (with Harold Rabinowitz) of *A Passion for Books: A Book Lover's Treasury of Stories, Essays, Humor, Lore, and Lists on Collecting, Reading, Borrowing, Lending, Caring For, and Appreciating Books,* published by Times Books in 1999, and lives with his wife and family in Cortlandt Manor, New York.

Acknowledgments

I would like to thank Paul Fargis of The Stonesong Press for suggesting that I compile this book, as well as John Michel of W. H. Freeman and Company for having the wisdom to publish it.

To Cecilia

Contents

Introduction

In the winter of 1675, Sir Isaac Newton wrote a letter to his fellow scientist Robert Hooke in which he used a phrase that has become as famous as almost any in science. "If I have seen further," he wrote, "it is by standing on the shoulders of Giants." Ever since, scientists have been standing on each other's shoulders in their ongoing efforts to glimpse even a small part of the answers to the mysteries that surround us. And one of the ways they lift themselves onto those august shoulders is by referring back to the work—and the words—of their predecessors. Thus, from the perspective of the scientist, gathering together the thoughts of scientists (as well as nonscientists) on the history, meaning, and practice of the discipline of science is a particularly appropriate endeavor.

And yet, *Science Says* is by no means intended only for the scientist, but—perhaps even more—for the nonscientist as well. We live in an increasingly chaotic and complex world—much of that chaos and complexity brought on by science—beginning perhaps with the splitting of the atom more than fifty years ago and the creation of a nuclear cloud that has hung over our heads ever since. As a result, science and those who eloquently speak and write about it have become central to how we view ourselves and the world. We look to interpreters of science for wisdom and answers, for insights into the nature of the universe and who we are, as well as for explanations of the everyday world in which we live. In fact, the words of scientists and science writers have taken the place once reserved for the pronouncements of philosophers, theologians, and literary figures. There is a great irony in this, particularly because, as James George Frazer put it in *The Golden Bough* more than a hundred years ago:

"The history of thought should warn us against concluding that because the scientific theory of the world is the best that has yet been formulated, it is necessarily complete and final. We must remember that at bottom the generalizations of science or, in common parlance, the laws of nature are merely hypotheses devised to explain that ever-shifting phantasmagoria of thought which we dignify with the high-sounding names of the world and the universe. In the last analysis magic, religion, and science are nothing but theories of thought."

Even so, a century later the nonscientific community—that is, the majority of humankind—has still not grasped this truth. As Louis Kronenberger said nearly fifty years ago, "Nominally a great age of scientific inquiry, ours has actually become an age of superstition about the infallibility of science; of almost mystical faith in its non-mystical methods. . . ." Or, as Thomas Stirton, one of my professors in college, put it more succinctly, "Science is our favorite modern superstition."

Paradoxically, although nonscientists consider science to be something akin to a religion, with immutable laws that cannot be questioned, scientists recognize its limitations and are always aware of the possibility—even probability—that beliefs must change to accommodate new information. They are eminently aware that the latest theories, however acceptable they may be at the moment, may well be over-turned tomorrow as new evidence comes to light. This understanding is a theme echoed throughout the pages that follow, an understanding that represents the essential spirit of science.

This spirit of science can be expressed in many ways, and this book is accordingly divided into fourteen subject areas covering everything from the meeting—and more than occasional clash—between science, spirit, and religion, to a variety of expressions of how scientists do science; from attempts to answer the question of where we've come from and where we're headed, to what we can know and what must remain a mystery; from the interface of science and society, to the place in science of imagination, intuition, curiosity, and creativity; and from efforts to define what science is and is not, to the thoughts of scientists on human-kind's place in the universe. In all, the following pages con-tain nearly a thousand expressions of the spirit of science. But perhaps none put it as eloquently as did Francis Albert Eley Crew in his essay, "The Meaning of Death":

"

"A few of the results of my activities as a scientist have become embedded in the very texture of the science I tried to serve—this is the immortality that every scientist hopes for. I have enjoyed the privilege, as a university teacher, of being in a position to influence the thought of many hundreds of young people and in them and in their lives I shall continue to live vicariously for a while. All the things I care for will continue for they will be served by those who come after me. I find great pleasure in the thought that those who stand on my shoulders will see much farther than I did in my time. What more could any man want?"

What indeed?

Rob Kaplan
Cortlandt Manor, NY
January 2000

Editor's Note

The quotations in this book are organized, within each subject area, alphabetically by the name of the person quoted. In those instances in which there is more than one quote from a particular person, the quotations are organized chronologically in three tiers. The first tier includes quotations for which the original date of utterance or writing is known. The second tier includes quotes for which this date is not known and is organized according to the publication dates of the books from which they were taken. The third and final tier includes those quotations that are attributed to the person quoted and for which there are no specific citations.

The form of the citations for the quotations is determined by the source, as follows:

- *A book by the person quoted:* Jacob Bronowski, *Science and Human Values* (1956)
- *A paper, essay, or article by the person quoted included in a collection of his or her work:* William James, "The Will to Believe," *The Will to Believe and Other Essays in Popular Philosophy* (1910)
- *A book written or edited by a person other than the one quoted:* Niels Bohr, in *The Cosmic Code* by Heinz R. Pagels (1982)

- *A paper, essay, or article by the person quoted included in an anthology edited by another person:* Erich Fromm, "The Creative Attitude," in *Creativity and Its Cultivation* edited by Harold H. Anderson (1959)
- *A paper, essay, or article by the person quoted included in a newspaper or periodical:* Albert Einstein, "Atomic War or Peace," *Atlantic Monthly,* November 1945
- *A newspaper or periodical:* Jack Horner, in *Time,* April 26, 1993
- *A speech or lecture by the person quoted:* Neil Armstrong, in a speech to Congress, September 16, 1969
- *A letter by the person quoted:* Marie Curie, in a letter to her brother, March 18, 1894.

Finally, in some instances quotations are cross-referenced to other quotations in the same section, such as when one statement is a response by its author to a statement by another author, or when several authors have made statements about the same subject. Such cross-references are indicated thus: [See J. Robert Oppenheimer].

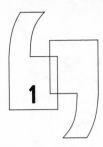

Science, Spirit, and Religion

The universe is merely a fleeting idea in God's mind—a pretty uncomfortable thought, particularly if you've just made a down payment on a house.

Woody Allen, *Getting Even* (1971)

For ourselves, we may take as a basic assumption, clear from a survey of particular cases, that natural things are some or all of them subject to change.

Aristotle, in *The Timeline Book of Science* by George Ochoa and Melinda Corey (1995)

If I get the impression that Nature itself makes the decisive choice [about] what possibility to realize, where quantum theory says that more than one outcome is possible, then I am ascribing personality to Nature, that is to something that is always everywhere. [An] omnipresent eternal personality which is omnipotent in taking the decisions that are left undetermined by physical law is exactly what in the language of religion is called God.

Frederik Jozef Belinfante, in *The World within the World*
by John D. Barrow (1988)

We have grasped the mystery of the atom and rejected the Sermon on the Mount.

General Omar Bradley, in a speech in
Boston, Massachusetts, November 10, 1948

To pursue science is not to disparage the things of the spirit. In fact, to pursue science rightly is to furnish the framework on which the spirit may rise.

Vannevar Bush, in a speech, October 5, 1953

This world, after all our science and sciences, is still a miracle; wonderful, inscrutable, magical and more, to whosoever will think of it.

Thomas Carlyle, *On Heroes, Hero-Worship
and the Heroic in History* (1841)

Gods are born and die, but the atom endures.

Alexander Chase, *Perspectives* (1966)

Theology is but the ignorance of natural causes reduced to a system. . . . [It] is a science that has for its object only things incomprehensible.

Paul Heinrich Dietrich d'Holbach, *Good Sense* (1772)

For my own part, I would as soon be descended from that heroic little monkey, who braved his dreaded enemy in order to save the life of his keeper; or from that old baboon, who, descending from the mountains, carried away in triumph his young comrade from a crowd of astonished dogs—as from a savage who delights to torture his enemies, offers up bloody sacrifices, practices infanticide without remorse, treats his wives like slaves, knows no decency, and is haunted by the grossest superstitions.

Charles Darwin, *On the Origin of Species* (1859)

I see no good reasons why the views given in this volume should shock the religious feelings of anyone.

Charles Darwin, *On the Origin of Species* (1859)

It is interesting to contemplate a tangled bank, clothed with many plants of many kinds, with birds singing on the bushes, with various insects flitting about, and with worms crawling through the damp earth, and to reflect that these elaborately constructed forms, so different from each other and dependent upon each other in so complex a manner, have all been produced by laws acting around us. These laws, taken in the largest sense, being Growth with Reproduction; Inheritance which is almost implied by reproduction; Variability from the indirect and direct action of the conditions of life, and from use and disuse; a Ratio of Increase so high as to lead to a Struggle for Life; and as a consequence to Natural Selection, entailing Divergence of Character and the extinction of less-improved forms. Thus, from the war of nature, from famine and death, that most exalted object which we are capable of conceiving, namely, the production of higher animals, directly follows. There is grandeur in this view of life, with its several powers, having been originally breathed by the Creator into a few forms or into one; and that, whilst this planet has gone cycling on according to the fixed law of gravity, from so simple a beginning endless forms most beautiful and most wonderful have been, and are being, evolved.

Charles Darwin, *On the Origin of Species* (1859)

It seems to me (rightly or wrongly) that direct arguments against Christianity and Theism hardly have any effect on the public; and that freedom of thought will best be promoted by that gradual enlightenment of human understanding which follows the progress of science. I have therefore always avoided writing about religion and have confined myself to science.

Charles Darwin, in a letter to Karl Marx, 1880, in *Ever Since Darwin* by Stephen Jay Gould (1978)

"

The objects which astronomy discloses afford subjects of sublime contemplation, and tend to elevate the soul above vicious passions and groveling pursuits.

> Thomas Dick, in *The Geography of the Heavens*
> by Elijah H. Burritt (1863)

Faith is a fine invention
When gentlemen can see,
But microscopes are prudent
In an emergency.

> Emily Dickinson, "Faith Is a Fine Invention,"
> *Poems, Second Series* (c. 1880)

Religion is the science of former times, dried out and turned to dogma; it is only the husk of an outdated scientific explanation.

> Roger Martin du Gard, *Jean Barois* (1949)

We have learned that matter is weird stuff. It is weird enough, so that it does not limit God's freedom to make it do what he pleases.

> Freeman J. Dyson, *Infinite in All Directions* (1988)

Life would be stunted and narrow if we could feel no significance in the world around us beyond that which can be weighed and measured with the tools of the physicist or

"

described by the metrical symbols of the mathematician.

> Sir Arthur Stanley Eddington,
> in *The World of Physics* by Arthur Beiser (1960)

You believe in a God who plays dice, and I in complete law and order in a world which objectively exists, and which I, in a wildly speculative way, am trying to capture. . . . Even the great initial success of the quantum theory does not make me believe in the fundamental dice game, although I am well aware that your younger colleagues interpret this as a consequence of senility.

> Albert Einstein, in a letter to Max Born, December 4, 1926
> [Usually quoted as "I cannot believe that God plays dice
> with the cosmos." See Stephen W. Hawking]

Everything is determined . . . by forces over which we have no control. It is determined for the insect as well as the star. Human beings, vegetables, or cosmic dust—we all dance to a mysterious tune, intoned in the distance by an invisible piper.

> Albert Einstein, in *The Saturday Evening Post*,
> October 26, 1929

The man who is thoroughly convinced of the universal operation of the law of causation cannot for a moment entertain the idea of a being who interferes in the course of events. . . . He has no use for the religion of fear and equally little for social or moral religion. A God who rewards and punishes is inconceivable to him for the simple reason that a man's actions are determined by necessity, external and internal, so that in God's eyes he cannot be responsible for the motions it undergoes. . . . A man's ethical behavior should be based effectually on sympathy, education, and social ties and needs; no religious basis is necessary. Man would indeed be in a poor way if he had to be restrained by fear of punishment and hope of reward after death.

Albert Einstein, "Religion and Science,"
The New York Times Magazine, November 9, 1930

I cannot imagine a God who rewards and punishes the objects of his creation, whose purposes are modeled after our own—a God, in short, who is but a reflection of human frailty. . . . It is enough for me to contemplate the mystery of conscious life perpetuating itself through all eternity, to reflect upon the marvelous structure of the universe which we can dimly perceive and try humbly to comprehend even an infinitesimal part of the intelligence manifested in Nature.

Albert Einstein, "My Credo," 1932, in *The Quotable Einstein*
edited by Alice Calaprice (1996)

A human being is part of the whole, called by us "Universe," a part limited in time and space. He experiences himself, his thoughts and feelings as something separated from the rest—a kind of optical delusion of his consciousness. This delusion is a kind of prison for us, restricting us to our personal desires and to affection for a few persons nearest to us. Our task must be to free ourselves from this prison by widening our circle of compassion to embrace all living creatures and the whole [of] nature in its beauty. Nobody is able to achieve this completely, but the striving for such achievement is in itself a part of the liberation and a foundation for inner security.

Albert Einstein, in a letter, March 4, 1950

Science without religion is lame, religion without science is blind.

Albert Einstein, "Science and Religion,"
Ideas and Opinions (1954)

My religion consists of a humble admiration of the illimitable superior spirit who reveals himself in the slight details we are able to perceive with our frail and feeble minds. That deeply emotional conviction of the presence of a superior reasoning power, which is revealed in the incomprehensible universe, forms my idea of God.

Albert Einstein, quoted in his obituary
in *The New York Times*, April 19, 1955

Coincidence is God's way of remaining anonymous.

Albert Einstein, in *The Word* by Noah ben Shea (1995)

I remain oppressed by the thought that the venture into space is meaningless unless it coincides with a certain interior expansion, an ever-growing universe within, to correspond with the far flight of the galaxies our telescopes follow from without.

Loren Eiseley, "The Inner Galaxy," *The Star Thrower* (1978)

The religion that is afraid of science dishonors God and commits suicide.

Ralph Waldo Emerson, Journal, 1857

The laws of nature are but the mathematical thoughts of God.

Euclid, in *A Mathematical Journey* by Stanley Gudder (1976)

Why is nature so nearly symmetrical? No one has any idea why. The only thing we might suggest is something like this: There is a gate in Japan, a gate in Neiko, which is sometimes called by the Japanese the most beautiful gate in Japan; it was built in a time when there was great influence from Chinese art. The gate is very elaborate, with lots of gables and beautiful carvings and lots of columns and dragon heads and princes carved into the pillars, and so on. But when one looks closely he sees that in the elaborate and complex design along one of the pillars, one of the small design elements is carved upside down; otherwise the thing is completely symmetrical. If one asks why this is, the story

"

is that it was carved upside down so that the gods will not be jealous of the perfection of man. So they purposely put the error in there, so that the gods would not be jealous and get angry with human beings.

We might like to turn the idea around and think that the true explanation of the near symmetry of nature is this: that God made the laws only nearly symmetrical so that we should not be jealous of His perfection.

Richard P. Feynman, *The Feynman Lectures on Physics,*
Volume One (1963)

A walk in the rainforest is a walk into the mind of God.

Birute M. F. Galdikas, *Reflections of Eden* (1995)

I, Galileo Galilei, son of the late Vincenzio Galilei of Florence, aged seventy years, being brought personally to judgment, and kneeling before you, Most Eminent and Most Reverend Lords Cardinals, General Inquisitors of the Universal Christian Commonwealth against heretical depravity, having before my eyes the Holy Gospels which I touch with my own hands, swear that I have always believed, and, with the help of God, will in future believe, every article which the Holy Catholic and Apostolic Church of Rome holds, teaches, and preaches. But because I have been enjoined, by the Holy Office, altogether to abandon the false opinion which maintains that the Sun is the center and immovable, and forbidden to hold, defend, or teach,

the said false doctrine in any manner . . . I am willing
to remove from the minds of your Eminences, and of
every Catholic Christian, this vehement suspicion rightly
entertained towards me, therefore, with a sincere heart and
unfeigned faith, I abjure, curse, and detest the said errors
and heresies, and generally every other error and sect
contrary to the said Holy Church; and I swear that I will
never more in future say, or assert anything, verbally or in
writing, which may give rise to a similar suspicion of me;
but that if I shall know any heretic, or anyone suspected of
heresy, I will denounce him to this Holy Office, or to the
Inquisitor and Ordinary of the place in which I may be. I
swear, moreover, and promise that I will fulfill and observe
fully all the penances which have been or shall be laid on
me by this Holy Office. But if it shall happen that I violate
any of my said promises, oaths, and protestations (which
God avert), I subject myself to all the pains and punishments
which have been decreed and promulgated by the sacred
canons and other general and particular constitutions against
delinquents of this description. So, may God help me, and
his Holy Gospels, which I touch with my own hands, I, the
above named Galileo Galilei, have abjured, sworn, promised,
and bound myself as above; and, in witness thereof, with
my own hand have subscribed this present writing of my
abjuration, which I have recited word for word.

Galileo Galilei, in *Galileo, His Life and Work*
by J. J. Fahie (1903)

"

Religion is a way of life and an attitude to the universe. It brings man into closer touch with the inner nature of reality. Statements of fact made in its name are untrue in detail, but often contain some truth at their core. Science is also a way of life and an attitude to the universe. It is concerned with everything but the nature of reality. Statements of fact made in its name are generally right in detail, but can only reveal the form and not the real nature of existence. The wise man regulates his conduct by the theories both of religion and science. But he regards these theories not as statements of ultimate fact, but as art forms.

> J. B. S. Haldane, "Science and Theology as Art Forms,"
> *Possible Worlds and Other Papers* (1927)

God not only plays dice. He also sometimes throws the dice where they cannot be seen.

> Stephen W. Hawking, in *Nature,* 1975 [See Albert Einstein]

People have always wanted answers to the big questions. Where did we come from? How did the world begin? What is the meaning and design behind it all? The creation accounts of the past now seem less credible. They have been replaced by a variety of superstitions, ranging from New Age to *Star Trek.* But real science can be far stranger than science fiction and much more satisfying.

> Stephen W. Hawking, *The Sunday Telegraph,* July 26, 1998

History warns us . . . that it is the customary fate of new truths to begin as heresies and to end as superstitions; and, as matters now stand, it is hardly rash to anticipate that, in another twenty years, the new generation, educated under the influences of the present day, will be in danger of accepting the main doctrines of the "Origin of Species," with as little reflection, and it may be with as little justification, as so many of our contemporaries, twenty years ago, rejected them.

T. H. Huxley, "The Coming of Age of 'The Origin of Species,'"
Darwinia: Essays (1896)

I asserted—and I repeat—that a man has no reason to be ashamed of having an ape for his grandfather. If there were an ancestor whom I should feel shame in recalling it would rather be a man—a man of restless and versatile intellect—who, not content with an equivocal success in his own sphere of activity, plunges into scientific questions with which he has no real acquaintance, only to obscure them by an aimless rhetoric, and distract the attention of his hearers from the real point at issue by eloquent digressions and skilled appeals to religious prejudice.

T. H. Huxley, responding to Bishop Samuel Wilberforce's
question, in *The Life and Letters of Thomas Henry Huxley*
by Leonard Huxley (1900) [See Bishop Samuel Wilberforce]

The God whom science recognizes must be a God of universal laws exclusively, a God who does a wholesale, not a retail business. He cannot accommodate his processes to the convenience of individuals.

William James, *The Varieties of Religious Experience* (1902)

Fundamental science is a universal good that all people must be able to cultivate in complete freedom from every form of international servitude or intellectual colonialism. Basic research must be free with regard to political and economic powers, which must cooperate in its development without impeding its creativity or subjugating it to their own ends. Like any other truth, scientific truth must render account only to itself and to the supreme truth that is God, creator of man and of all things.

Pope John Paul II (Karol Wojtyla), in *Science,* March 1980

Science can purify religion from error and superstition. Religion can purify science from idolatry and false absolutes.

Pope John Paul II (Karol Wojtyla),
in *Galileo: A Life* by James Reston (1994)

Science investigates; religion interprets. Science gives man knowledge which is power; religion gives man wisdom which is control.

Martin Luther King, Jr., *Strength to Love* (1963)

Electronic calculators can solve problems which the
man who made them cannot solve; but no government-
subsidized commission of engineers and physicists could
create a worm.

Joseph Wood Krutch, "March,"
The Twelve Seasons (1949)

The search for meaning is not limited to science: it is
constant and continuous—all of us engage in it during all
our waking hours; the search continues even in our dreams.
There are many ways of finding meaning, and there are no
absolute boundaries separating them. One can find meaning
in poetry as well as in science; in the contemplations of a
flower as well as in the grasp of an equation. We can be
filled with wonder as we stand under the majestic dome of
the night sky and see the myriad lights that twinkle and
shine in its seemingly infinite depths. We can also be filled
with awe as we behold the meaning of the formulae that
define the propagation of light in space, the formation
of galaxies, the synthesis of chemical elements, and the
relation of energy, mass and velocity in the physical
universe. The mystical perception of oneness and the
religious intuition of a Divine intelligence are as much a
construction of meaning as the postulation of the universal
law of gravitation.

Ervin Laszlo

It is God who is the ultimate reason of things, and the knowledge of God is no less the beginning of science than his essence and will are the beginning of beings.

> Gottfried Wilhelm Leibnitz, *Letter on a General Principle Useful in Explaining the Laws of Nature* (1687)

The effort to reconcile science and religion is almost always made, not by theologians, but by scientists unable to shake off altogether the piety absorbed with their mother's milk.

> H. L. Mencken, *Minority Report: H. L. Mencken's Notebook* (1956)

The universe was dictated but not signed.

> Christopher Morley, in *Physically Speaking* edited by C. C. Gaither and A. E. Cavazos-Gaither (1997)

The next great task of science is to create a religion for mankind.

> John Morley, in *A Dictionary of Scientific Quotations* by Alan L. Mackay (1991)

What hath God wrought.

> Samuel F. B. Morse, first message sent by telegraph, a quote from the Bible (Numbers 23:23), May 24, 1844

No scientific explanation of the existence of the universe and the flesh-and-blood men who do the explaining has superseded the first four words of the Bible. Science does not tell *why* we and the universe about us exist at all. It has cautiously offered to explain *how* the world arrived at its present physical state; and over these details men have quarreled. For many centuries after their promulgation . . . the opening verses of the first of the five books of Moses gave what the wisest men of those times considered the most reasonable and hence the most scientific account of creation.

Forest Ray Moulton and Justus J. Schifferes,
The Autobiography of Science (Second Edition) (1960)

The Church welcomes technological progress and receives it with love, for it is an indubitable fact that technological progress comes from God and, therefore, can and must lead to Him.

Pope Pius XII, Christmas Message, 1953

For everyone, as I think, must see that astronomy compels the soul to look upwards and leads us from this world to another.

Plato, *The Republic* (5th–4th century B.C.)

Whenever it is possible to find out the cause of what is happening, one should not have recourse to the gods.

Polybius, in *The Theory of the Mixed Constitution
in Antiquity* by K. Van Fritz (1954)

"

All are but parts of one stupendous whole,
Whose body Nature is, and God the soul.

Alexander Pope, *An Essay on Man* (1733–34)

Educators may bring upon themselves unnecessary travail
by taking a tactless and unjustifiable position about the
relation between scientific and religious narratives. We see
this, of course, in the conflict concerning creation science.
Some educators representing, as they think, the conscience
of science act much like those legislators who in 1925
prohibited by law the teaching of evolution in Tennessee. In
that case, anti-evolutionists were fearful that a scientific idea
would undermine religious belief. Today, pro-evolutionists
are fearful that a religious idea will undermine scientific
belief. The former had insufficient confidence in religion;
the latter insufficient confidence in science. The point is
that profound but contradictory ideas may exist side by
side, if they are constructed from different materials and
methods and have different purposes. Each tells us something
important about where we stand in the universe, and it is
foolish to insist that they must despise each other.

Neil Postman, *The End of Education* (1995)

[It is] the handwriting of God.

Joel Primack, on the night sky, in *Newsweek*, May 4, 1992

It is only in science, I find, that we can get outside ourselves. It's realistic, and to a great degree verifiable, and it has this tremendous stage on which it plays. I have the same feeling—to a certain degree—about some religious expressions . . . but only to a certain degree. For me, the proper study of mankind is science, which also means that the proper study of mankind is man.

I. I. Rabi, in *Experiencing Science* by Jeremy Bernstein (1978)

To be happy in this world, especially when youth is past, it is necessary to feel oneself not merely an isolated individual whose day will soon be over, but part of the stream of life flowing on from the first germ to the remote and unknown future.

Bertrand Russell, *The Conquest of Happiness* (1930)

Those who cavalierly reject the Theory of Evolution as not being adequately supported by facts, seem to forget that their own theory is supported by no facts at all.

Herbert Spencer, "The Development Hypothesis" (1852), *Essays: Scientific, Political & Speculative* (1966)

It shall be unlawful for any teacher in any of the universities, normal and all other public schools of the state which are supported in whole or in part by the public school funds of the state, to teach any theory that denies the story of the divine creation of man as taught in the Bible, and to teach

"

instead that man has descended from a lower order of animals.

State of Tennessee Statute, 1925

Formerly, when religion was strong and science weak, men mistook magic for medicine; now, when science is strong and religion weak, men mistake medicine for magic.

Thomas Szasz, *The Second Sin* (1973)

Mystics always hope that science will some day overtake them.

Booth Tarkington, *Looking Forward* (1926)

Nature is full of genius, full of the divinity; so that not a snowflake escapes its fashioning hand.

Henry David Thoreau, Journal, January 5, 1856

Some say God is living there [in space]. I was looking around very attentively, but I did not see anyone there. I did not detect either angels or gods. . . . I don't believe in God. I believe in man—his strength, his possibilities, his reason.

Gherman Titov, Soviet cosmonaut,
in *The Seattle Daily Times*, May 7, 1962

I believe that our Heavenly Father invented man because he was disappointed in the monkey.

Mark Twain, autobiographical dictation, November 24, 1906

Religious feeling is as much a verity as any other part of human consciousness; and against it, on the subjective side, the waves of science beat in vain.

John Tyndall, "Professor Virchow and Evolution," *Fragments of Science for Unscientific People* (1871)

The world was created on 22nd October, 4004 B.C. at 6 o'clock in the evening.

James Ussher, Bishop of Armagh, *Chronologia Sacra* (1660)

"One sacred memory from childhood is perhaps the best education," said Feodor Dostoevski. I believe that, and I hope that many earthling children will respond to the first human footprint on the moon as a sacred thing. We need sacred things.

Kurt Vonnegut, Jr., *Wampeters, Foma, and Granfalloons* (1974)

We are the products of editing, rather than authorship.

George Wald, "The Origin of Optical Activity," *Annals of the New York Academy of Science,* 1975

The greatest spiritual revolutionary in Western history, Saint Francis, proposed what he thought was an alternative Christian view of nature and man's relation to it: he tried to substitute the idea of the equality of all creatures, including man, for the idea of man's limitless rule of creation. He failed. Both our present science and our present technology are so tinctured with orthodox Christian arrogance toward nature that no solution for our ecologic crisis can be expected from them alone. Since the roots of our trouble are so largely religious, the remedy must also be essentially religious, whether we call it that or not. We must rethink and refeel our nature and destiny. The profoundly religious, but heretical, sense of primitive Franciscans for the spiritual autonomy of all parts of nature may point a direction.

> Lynn T. White, Jr., "The Historical Roots of
> Our Ecologic Crisis," *Science,* 1967

Religion will not regain its old power until it can face change in the same spirit as does science. Its principles may be eternal, but the expression of those principles requires continual development.

> Alfred North Whitehead, *Science and the Modern World* (1925)

If anyone were to be willing to trace his descent through an ape as his *grandfather,* would he be willing to trace his descent similarly on the side of his *grandmother*?

> Bishop Samuel Wilberforce, in a speech to the British
> Association for the Advancement of Science, 1860
> [See T. H. Huxley]

Religions die when they are proved to be true. Science is the record of dead religions.

> Oscar Wilde, *Phrases and Philosophies
> for the Use of the Young* (1894)

2

Chaos and Order

In plain words, Chaos was the law of nature; Order was the dream of man.

Henry Adams, *The Education of Henry Adams* (1918)

The notion that the "balance of nature" is delicately poised and easily upset is nonsense. Nature is extraordinarily tough and resilient, interlaced with checks and balances, with an astonishing capacity for recovering from disturbances in equilibrium. The formula for survival is not power; it is symbiosis.

Eric Ashby, in *Encounter*, March 1976

The more man inquires into the laws which regulate the material universe, the more he is convinced that all its varied forms arise from the action of a few simple principles. These principles themselves converge, with accelerating force, towards some still more comprehensive law to which all matter seems to be submitted. Simple as that law may possibly be, it must be remembered that it is only one amongst an infinite number of simple laws: that each of these laws has consequences at least as extensive as the existing one, and therefore that the Creator who selected the present law must have foreseen the consequences of all other laws.

> Charles Babbage, in *Theories of Everything*
> by John D. Barrow (1991)

The universe is simmering down, like a giant stew left to cook for four billion years. Sooner or later we won't be able to tell the carrots from the onions.

> Arthur Bloch, in *The World within the World*
> by John D. Barrow (1988)

All things begin in Order, so shall they end, and so shall they begin again, according to the Ordainer of Order, and the mystical mathematicks of the City of Heaven.

> Sir Thomas Browne, *Hydriotaphia,*
> *Urn Burial and the Garden of Cyrus* (1896)

. . . all the work of the crystallographers serves only to demonstrate that there is only variety everywhere where they suppose uniformity . . . that in nature there is nothing absolute, nothing perfectly regular.

> Georges Leclarc Buffon, *Histoire Naturelle*
> *des Minéraux* (1783–88)

When we observe nature we see what we want to see, according to what we believe we know about it at the time. Nature is disordered, powerful and chaotic, and through fear of the chaos we impose system on it. We abhor complexity, and seek to simplify things whenever we can by whatever means we have at hand. We need to have an overall explanation of what the universe is and how it functions. In order to achieve this overall view we develop explanatory theories which will give structure to natural phenomena: we classify nature into a coherent system which appears to do what we say it does.

James Burke, *The Day the Universe Changed* (1985)

When people talk of atoms obeying fixed Laws, they are either ascribing some kind of intelligence and free will to atoms or they are talking nonsense.

Samuel Butler, *Samuel Butler's Notebooks*
edited by Geoffrey Keyner and Brian Hill (1951)

If order appeals to the intellect, then disorder titillates the imagination.

Paul Claudel, in *Structure of Non-crystalline Materials*
edited by P. H. Gaskell (1977)

. . . the rules of clockwork might apply to familiar objects such as snookerballs, but when it comes to atoms, the rules are those of roulette.

Paul Davies, *God and the New Physics* (1983)

"

The universe contains vastly more order than Earth-life
could ever demand. All those distant galaxies, irrelevant for
our existence, seem as equally well ordered as our own.

> Paul Davies, in *The Quickening Universe*
> by Eugene F. Mallove (1987)

. . . the universe is but a watch on a larger scale; all its
motions depending on determined laws and mutual relation
of its parts.

> Bernard de Fontenelle, *Conversations*
> *on the Plurality of Worlds* (1803)

Everything existing in the Universe is the fruit of chance
and necessity.

> Democritos, in *A Dictionary of Scientific Quotations*
> by Alan L. Mackay (1991)

Law rules throughout existence, a Law which is not
intelligent but Intelligence.

> Ralph Waldo Emerson, "Fate," *The Conduct of Life* (1860)

For since the fabric of the universe is most perfect and the
work of a most wise Creator, nothing at all takes place in
the universe in which some rule of maximum or minimum
does not appear.

> Leonhard Euler, in *Mathematical Thought from Ancient*
> *to Modern Times* by Morris Kline (1972)

Unfortunately, non-chaotic systems are very nearly as scarce as hen's teeth, despite the fact that our physical understanding of nature is largely based upon their study. . . . For centuries, randomness has been deemed a useful, but subservient citizen in a deterministic universe. Algorithmic complexity theory and nonlinear dynamics together establish the fact that determinism actually reigns over a quite finite domain; outside this small haven of order lies a largely uncharted, vast wasteland of chaos where determinism has faded into an ephemeral memory of existence theorems and only randomness survives.

> Joseph Ford, "How Random Is a Coin Toss?"
> *Physics Today,* April 1983

Nature goes her own way, and all that to us seems an exception is really according to order.

> Johann Wolfgang von Goethe, December 8, 1824,
> in *Conversations with Goethe* by Johann Peter Eckermann

The progress of the human race in understanding the universe has established a small corner of order in an increasingly disordered universe.

> Stephen W. Hawking, *A Brief History of Time* (1988)

The whole history of science has been the gradual realization that events do not happen in an arbitrary manner, but that

"

they reflect a certain underlying order, which may or may not be divinely inspired.

Stephen W. Hawking, *A Brief History of Time* (1988)

I am much occupied with the investigation of physical causes. My aim in this is to show that the celestial machine is to be likened not to a divine organism, but rather a clockwork. . . .

Johannes Kepler, in *Astronomy: The Evolving Universe* by Michael Zeilik (1982)

Confusion evolves into order spontaneously. What God really said was, "Let there be chaos."

Rosario M. Levins, in *Hierarchy Theory* edited by H. H. Pattee (1973)

Whence is it that nature does nothing in vain; and whence arises all that order and beauty which we see in the world?

Sir Isaac Newton, *Opticks* (1704)

Instead of finding an absolute universal law at the bottom of existence, they may find an endless regress of laws, or even worse, total confusion and lawlessness—an outlaw universe.

Heinz R. Pagels, *Perfect Symmetry* (1985)

The fact that the universe is governed by simple natural laws is remarkable, profound and on the face of it absurd. How can the vast variety in nature, the multitude of things and processes all be subject to a few simple, universal laws?

Heinz R. Pagels, *Perfect Symmetry* (1985)

We no longer pretend to be able to grasp reality in a physical theory; we see in it rather an analytic or geometric mold useful and fertile for a tentative representation of phenomena, no longer believing that the agreement of a theory with experience demonstrates that the theory expresses the reality of things. Such statements have sometimes seemed discouraging; we ought rather to marvel that, with representations of things more or less distant and discolored, the human spirit has been able to find its way through the chaos of so many phenomena and to derive from scientific knowledge the ideas of beauty and harmony. It is no paradox to say that science puts order, at least tentative order, into nature.

Emile Picard, in *The Modern Aspect of Mathematics* by Lucienne Felix (1960)

How do we discover the individual laws of Physics, and what is their nature? It should be remarked, to begin with, that we have no right to assume that any physical law exists, or if they have existed up to now, that they will continue to exist in a similar manner in the future. It is

"

perfectly conceivable that one fine day Nature should cause an unexpected event to occur which would baffle us all; and if this were to happen we would be powerless to make any objection, even if the result would be that, in spite of our endeavors, we should fail to introduce order into the resulting confusion. In such an event, the only course open to science would be to declare itself bankrupt. For this reason, science is compelled to begin by the general assumption that a general rule of law dominates throughout Nature. . . .

> Max Planck, *The Universe in the*
> *Light of Modern Physics* (1931)

Order is Heav'ns's first law.

> Alexander Pope, "An Essay on Man" (1733–34)

The irreversibility [of time] is the mechanism that brings order out of chaos.

> Ilya Prigogine, in *The Big Bang Never Happened*
> by Eric J. Lerner (1991)

There is . . . one supremely important law which is only statistical; this is the second law of thermodynamics. It states, roughly speaking, that the world is growing continuously more disorderly.

> Bertrand Russell, "Scientific Metaphysics,"
> *The Scientific Outlook* (1931)

Nothing comes to pass in nature, which can be set down
to a flaw therein; for nature is always the same, and
everywhere one and the same in her efficacy and power of
action; that is, nature's laws and ordinances, whereby all
things come to pass and change from one form to another,
are everywhere and always the same; so that there should
be one and the same method of understanding the nature of
all things whatsoever, namely, through nature's universal
laws and rules.

Baruch Spinoza, The Ethics (1677)

Scientists speak of the Law of Inertia or the Second Law of
Thermodynamics as if some great legislature in the sky once
met and set down rules to govern the universe.

Victor J. Stenger, Not by Design (1988)

. . . chaos is "lawless behavior governed entirely by law."

Ian Stewart, Does God Play Dice? (1990)

How to separate dice from design—that is the major
question facing evolutionary biology today.

Tyler Volk, in The Sciences, May/June 1990

. . . it would be very singular that all nature, all the planets,
should obey eternal laws, and that there should be a little

”

animal five feet high who, in contempt of these laws, could act as he pleased, solely according to his caprice.

> Voltaire (François-Marie Arouet), "Ignorant Philosophers,"
> *The Best Known Works of Voltaire* (1940)

Everything in space obeys the laws of physics. If you know these laws, space will treat you kindly.

> Wernher von Braun, in *Time,* February 17, 1958

Time is what prevents everything from happening at once.

> John Archibald Wheeler, in
> *The American Journal of Physics,* 1978

Nature, so far as it is the object of scientific research, is a collection of facts governed by *laws:* our knowledge of nature is our knowledge of laws.

> William Whewell, *Astronomy and General Physics,*
> *considered with reference to Natural Theology* (1834)

Both science and art have to do with ordered complexity.

> Lancelot Law White, in *The Griffin,* 1957

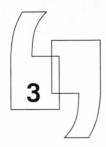

3

What Science Is . . . and Is Not

To try to write a grand cosmical drama leads necessarily to myth. To try to let knowledge substitute [for] ignorance in increasingly larger regions of space and time is science.

> Hannes Alfvén, in *The Big Bang Never Happened*
> by Eric J. Lerner (1991)

The quick harvest of applied science is the useable process, the medicine, the machine. The shy fruit of pure science is Understanding.

> Lincoln Barnett, on Einstein's completion of a mathematical
> formula for the United Field Theory, in *Life,* January 9, 1950

It is wrong to think that the task of physics is to find out how Nature is. Physics concerns what we can say about Nature.

> Niels Bohr, in *The Cosmic Code* by Heinz R. Pagels (1982)

Science doesn't deal with facts; indeed . . . fact is an emotion-loaded word for which there is little place in scientific debate.

> Sir Herman Bondi, in *Nature*, 1977

I am now convinced that theoretical physics is actual philosophy.

> Max Born, *Autobiography*

Art upsets, science reassures.

> Georges Braque, *Pensées sur l'Art*

Science knows only one commandment: contribute to science.

> Bertolt Brecht, *The Life of Galileo* (1972)

Man masters nature not by force but by understanding. That is why science has succeeded where magic failed: because it has looked for no spell to cast on nature.

> Jacob Bronowski, "The Creative Mind,"
> *Science and Human Values* (1956)

Man is unique not because he does science, and he is unique not because he does art, but because science and art equally are expressions of his marvelous plasticity of mind.

> Jacob Bronowski, *The Ascent of Man* (1973)

I find it [science] analytic, pretentious, superficial, largely because it does not address itself to dreams, chance, laughter, feelings or paradox—all the things I love the most.

Luis Buñuel, *My Last Sigh* (1983)

Science has a simple faith, which transcends utility. It is the faith that it is the privilege of man to learn to understand, and that this is his mission.

Vannevar Bush, *Science Is Not Enough* (1967)

We live in a scientific age, yet we assume that knowledge of science is the prerogative of only a small number of human beings, isolated and priestlike in their laboratories. This is not true. The materials of science are the materials of life itself. Science is part of the reality of living; it is the way, the how and the why for everything in our experience.

Rachel Carson, in a speech, 1952

Remember, then, that it [science] is the guide of action; that the truth which it arrives at is not that which we can ideally contemplate without error, but that which we may act upon without fear; and you cannot fail to see that scientific thought is not an accompaniment or condition of human progress, but human progress itself.

William Kingdon Clifford,
Aims and Instruments of Scientific Thought (1872)

"

Poetry is not the proper antithesis to prose, but to science. . . . The proper and immediate object of science is the acquirement, or communication, of truth; the proper and immediate object of poetry is the communication of immediate pleasure.

Samuel Taylor Coleridge, *Definitions of Poetry* (1811)

The first man of science was he who looked into a thing, not to learn whether it furnished him with food, or shelter, or weapons, or tools, or armaments, or playwiths but who sought to know it for the gratification of knowing.

Samuel Taylor Coleridge, in *A Dictionary of Scientific Quotations* by Alan L. Mackay (1991)

To say that science is logical is like saying that a painting is paint.

Leon Cooper, in *In the Palace of Memory* by George Johnson (1991)

Science is the meeting place of two kinds of poetry: the poetry of thought and the poetry of action.

George Agostinho da Silva, in *An Imagined World* by June Goodfield (1981)

There is an order which regulates our progress. Every science develops after a certain number of preceding sciences have been developed, and only then; it has to await its turn to burst its shell.

> Bernard de Fontenelle, *Préface des Éléments de la Géométrie de l'Infinie, Oeuvres* (1790)

Scientific principles and laws do not lie on the surface of nature. They are hidden, and must be wrested from nature by an active and elaborate technique of inquiry.

> John Dewey, *Reconstruction in Philosophy* (1920)

Every science begins as philosophy and ends as art.

> Will Durant, *The Story of Philosophy* (1926)

After a certain high level of technical skill is achieved, science and art tend to coalesce in esthetics, plasticity, and form. The greatest scientists are always artists as well.

> Albert Einstein, 1923, in the *Durham Morning Herald,* August 21, 1955

The grand aim of all science is to cover the greatest number of empirical facts by logical deduction from the smallest number of hypotheses or axioms.

> Albert Einstein, in *Life,* January 9, 1950

Science can only state what is, not what should be.

> Albert Einstein, *Out of My Later Years* (1950)

Science is the attempt to make the chaotic diversity of our sense-experience correspond to a logically uniform system of thought.

> Albert Einstein, *Out of My Later Years* (1950)

The whole of science is nothing more than a refinement of everyday thinking. It is for this reason that the critical thinking of the physicist cannot possibly be restricted by the examination of the concepts of his own specific field. He cannot proceed without considering critically a much more difficult problem, the problem of analyzing the nature of everyday thinking.

> Albert Einstein, *Out of My Later Years* (1950)

. . . all our science, measured against reality, is primitive and childlike—and yet it is the most precious thing we have.

> Albert Einstein, in "Glimpses of Einstein—a photo essay,"
> *The Physics Teacher,* April 1974

While natural science up to the end of the last century was predominantly a *collecting* science, a science of finished things, in our century it is essentially a *classifying* science, a science of processes, of the origin and development of these things and of the interconnection which binds all these processes into one great whole.

> Friedrich Engels, speaking of the 18th and
> 19th centuries, *Ludwig Fuerbach* (1886)

Science is a long history of learning how not to fool ourselves.

Richard P. Feynman

Far from being a difficult and inaccessible science, Astronomy is the science which concerns us the most, the one most necessary for our general instruction, and at the same time the one which offers for our study the greatest charms and keeps in reserve the highest enjoyments. We cannot be indifferent to it, for it alone teaches us where we are and what we are; and, moreover, it need not bristle with figures, as some severe *savants* would wish us to believe. The algebraical formulae are merely scaffoldings analogous to those which are used to construct an admirably designed palace. The figures drop off, and the palace of Urania shines in the azure, displaying to our wondering eyes all its grandeur and all its magnificence.

Camille Flammarion, *Popular Astronomy:
A General Description of the Heavens* (1894)

Science, in the very act of solving problems, creates more of them.

Abraham Flexner, *Universities* (1930)

A science is any discipline in which the fool of this generation can go beyond the point reached by the genius of the last generation.

Max Gluckman, *Politics, Law and Ritual* (1965)

"

Grey and ashen, my friend, is every science. And only the golden tree of life is green.

> Johann Wolfgang von Goethe, *Faust* (1808)

Science is all those things which are confirmed to such a degree that it would be unreasonable to withhold one's provisional consent.

> Stephen Jay Gould, in a lecture on evolution,
> at Cambridge University, 1984

Science is not to be regarded merely as a storehouse of facts to be used for material purposes, but as one of the great human endeavors to be ranked with arts and religion as the guide and expression of man's fearless quest for truth.

> Sir Richard Arman Gregory, in *A Dictionary of Scientific Quotations* by Alan L. Mackay (1991)

Natural science does not simply describe and explain nature; it is part of the interplay between nature and ourselves; it describes nature as exposed to our method of questioning.

> Werner Heisenberg, *Physics and Philosophy* (1959)

Whoever, in the pursuit of science, seeks after immediate practical utility may rest assured that he will seek in vain. All that science can achieve is a perfect knowledge and a perfect understanding of the action of natural and moral forces.

> Hermann Ludwig Ferdinand von Helmholtz,
> academic discourse, Heidelberg, 1862

"

In all human activities, it is not ideas or machines that dominate; it is people. I have heard people speak of "the effect of personality on science." But this is a backward thought. Rather, we should talk about the effect of science on personalities. Science is not the dispassionate analysis of impartial data. It is the human, and thus passionate, exercise of skill and sense on such data. Science is not an exercise in objectivity, but, more accurately, an exercise in which objectivity is prized.

Philip Hilts, *Scientific Temperaments: Three Lives in Contemporary Science* (1982)

Science is the knowledge of Consequences, and dependence of one fact upon another.

Thomas Hobbes, *Leviathan* (1651)

Science is a first-rate piece of furniture for a man's upper chamber, if he has common sense on the ground floor.

Oliver Wendell Holmes, Sr., *The Poet at the Breakfast-Table* (1872)

Science is the topography of ignorance.

Oliver Wendell Holmes, Sr., *Medical Essays* (1883)

Equipped with his five senses, man explores the universe around him and calls the adventure Science.

Edwin Powell Hubble, *The Nature of Science* (1954)

"

Applied Science is a conjuror, whose bottomless hat yields impartially the softest of Angora rabbits and the most petrifying of Medusas.

Aldous Huxley, *Tomorrow and Tomorrow and Tomorrow* (1956)

All our scientific and philosophic ideals are altars to unknown gods.

William James, *The Dilemma of Determinism* (1884)

Science can tell us what exists; but to compare the *worths,* both of what exists and of what does not exist, we must consult not science, but . . . our heart.

William James, "The Will to Believe," *The Will to Believe and Other Essays in Popular Philosophy* (1910)

Science . . . is not a description of the physical world, but a description of how the world interacts with the mind—and how experience is translated into the structures we call memories.

George Johnson, *In the Palace of Memory* (1991)

What the founders of modern science . . . had to do, was not to criticize and to combat certain faulty theories, and to correct or to replace them by better ones. They had to do something quite different. They had to destroy one world and replace it by another. They had to reshape the framework of our intellect itself, to restate and to reform its concepts, to evolve a new approach to Being, a new concept of knowledge, a new concept of science—and even to replace a pretty natural approach, that of common sense, by another which is not natural at all.

> Alexandre Koyré, "Galileo and Plato,"
> *Journal of the History of Ideas,* 1943

It is absurd to deny the role of fantasy in even the strictest science.

> V. I. Lenin, *Polnoe Sobranie Sochinenii*

Knowledge is a matter of science, and no dishonesty or conceit whatsoever is permissible. What is required is definitely the reverse—honesty and modesty.

> Mao Tse-Tung, "On Practice," *Quotations from
> Chairman Mao Tse-Tung* (1967)

Science, at bottom, is really anti-intellectual. It always distrusts pure reason, and demands the production of objective fact.

> H. L. Mencken, *Minority Report:
> H. L. Mencken's Notebook* (1956)

Most institutions demand unqualified faith; but the institution of science makes skepticism a virtue.

> Robert K. Merton, *Social Theory and Social Structure* (1962)

"

We must not wait for favors from Nature; our task is to wrest them from her.

Ivan Vladimirovich Michurin,
A Short Dictionary of Philosophy (1955)

In science, self-satisfaction is death. Personal self-satisfaction is the death of the scientist. Collective self-satisfaction is the death of the research. It is restlessness, anxiety, dissatisfaction, agony of mind that nourish science.

Jacques Monod, in *New Scientist,* June 17, 1976

What is a scientist? . . . We give the name scientist to the type of man who has felt experiment to be a means guiding him to search out the deep truth of life, to lift a veil from its fascinating secrets, and who, in this pursuit, has felt arising within him a love for the mysteries of nature, so passionate as to annihilate the thought of himself.

Maria Montessori, *The Montessori Method* (1964)

It is the desire for explanations that are at once systematic and controllable by factual evidence that generates science; and it is the organization and classification of knowledge on the basis of explanatory principles that is the distinctive goal of the sciences.

Ernest Nagel, *The Structure of Science* (1961)

The greatest of the changes that science has brought is the acuity of change; the greatest novelty the extent of novelty.

J. Robert Oppenheimer, in a lecture, 1953

Science will never be able to reduce the value of a sunset to arithmetic. Nor can it reduce friendship to formula. Laughter and love, pain and loneliness, the challenge of accomplishment in living, and the depth of insight into beauty and truth: these will always surpass the scientific mastery of nature.

Louis Orr, in a speech to the American
Medical Association, June 6, 1960

We must renounce the hope of representing the physical world by referring natural phenomena to a mechanics of atoms. "But"—I hear you say—"but what will we have left to give us a picture of reality if we abandon atoms?" To this I reply: "Thou shalt not take unto thee any graven images, or any likeness of anything." Our task is not to see the world through a dark and distorted mirror, but directly, so far as the nature of our minds permits. The task of science is to discern relationships among realities. . . .

Friedrich Wilhelm Ostwald,
in *Quantum Reality* by Nick Herbert (1985)

Science provides a vision of reality seen from the perspective of reason, a perspective that sees the vast order of the

"

universe, living and nonliving matter, as a material system governed by rules that can be known by the human mind. It is a powerful vision, formal and austere but strangely silent about many of the questions that deeply concern us. Science shows us what exists but not what to do about it.

Heinz R. Pagels, *The Dreams of Reason* (1988)

It is a fraud of the Christian system to call the sciences human invention; it is only the application of them that is human. Every science has for its basis a system of principles as fixed and unalterable as those by which the universe is regulated and governed. Man cannot make principles; he can only discover them.

Thomas Paine, *The Age of Reason* (1794–95)

Above, far above, the prejudices and passions of men soar the laws of nature. Eternal and immutable, they are the expression of the creative power; they represent what is, what must be, what otherwise could not be. Man can come to understand them: he is incapable of changing them. From the infinitely great down to the infinitely small, all things are subject to them. The sun and the planets follow the laws discovered by Newton and Laplace, just as the atoms in their combinations follow the laws of chemistry, as living creatures follow the laws of biology. It is only the imperfections of the human mind which multiply the divisions of the sciences, separating astronomy from physics or chemistry, the natural sciences from the social sciences. In essence, science is one. It is none other than the truth.

> Vilfredo Pareto, *Cours d'economie Politique* (1896–97)

. . . there does not exist a category of science to which one can give the name applied science. There are science and the applications of science, bound together as the fruit to the tree which bears it.

> Louis Pasteur, in *Revue Scientifique,* 1871

The most remarkable discovery made by scientists is science itself.

> Gerard Piel, in *Magic, Science and Civilization*
> by Jacob Bronowki (1978)

Traditional scientific method has always been at the very *best,* 20-20 hindsight. It's good for seeing where you've been.

> Robert M. Pirsig, *Zen and the Art of*
> *Motorcycle Maintenance* (1974)

An experiment is a question which science poses to Nature, and a measurement is the recording of Nature's answer.

> Max Planck, *Scientific Autobiography*
> *and Other Papers* (1949)

"

Science is built up of facts, as a house is built of stones; but an accumulation of facts is no more a science than a heap of stones is a house.

> Henri Poincaré, *Science and Hypothesis* (1905)

The Republic of Science is a Society of Explorers. Such a society strives towards an unknown future, which it believes to be accessible and worth achieving. In the case of scientists, the explorers strive towards a hidden reality, for the sake of intellectual satisfaction. And as they satisfy themselves, they enlighten all men and are thus helping society to fulfill its obligation toward intellectual self-improvement.

> Michael Polanyi, *Minerva*, 1962

Science is not a system of certain, or well-established, statements; nor is it a system which steadily advances towards a state of finality. . . . And our guesses are guided by the unscientific, the metaphysical (though biologically explicable) faith in laws, in regularities which we can uncover—discover. Like Bacon, we might describe our own contemporary science—"the method of reasoning which men now ordinarily apply to nature"—as consisting of "anticipations, rash and premature" and as "prejudices."

> Sir Karl Popper, *The Logic of Scientific Discovery* (1959)

"The scientific method," Thomas Henry Huxley once wrote, "is nothing but the normal working of the human mind." That is to say, when the mind is working; that is to say further, when it is engaged in correcting its mistakes. Taking this point of view, we may conclude that science is not physics, biology, or chemistry—is not even a "subject"—but a moral imperative drawn from a larger narrative whose purpose is to give perspective, balance, and humility to learning.

Neil Postman, *The End of Education* (1995)

Science is not just the fruit of the tree of knowledge, it is the tree itself.

Derek J. de Solla Price, in a lecture given in London, 1964

The work of science is to substitute facts for appearances, and demonstrations for impressions.

John Ruskin, *The Stones of Venice* (1851–53)

Science is a way of thinking much more than it is a body of knowledge. Its goal is to find out how the world works, to seek what regularities there may be, to penetrate to the connections of things—from subatomic particles, which may be the constituents of all matter, to living organisms, the human social community, and thence to the cosmos as a whole.

Carl Sagan, *Broca's Brain: Reflections on the Romance of Science* (1979)

"

Science is nothing but developed perception, interpreted intent, common sense rounded out and minutely articulated.

> George Santayana, *The Life of Reason* (1905–06)

Science is always wrong. It never solves a problem without creating ten more.

> George Bernard Shaw, in *The Timeline Book of Science* by George Ochoa and Melinda Corey (1995)

What distinguishes science from every other form of human intellectual activity is that it disciplines speculation with facts. Theory and data are the two blades of the scissors. But the metaphor is not quite right, for the blades are not symmetric. When theories and facts are in conflict, the theories must yield.

> Herbert A. Simon, in *The State of Economic Science* edited by Werner Sichel (1989)

Science is the great antidote to the poison of enthusiasm and superstition.

> Adam Smith, *The Wealth of Nations* (1776)

Let science neither be a crown to put proudly on your head nor an axe to chop wood.

> Talmud

If there ever was a misnomer, it is "exact science." Science
has always been full of mistakes. The present day is no
exception. And our mistakes are good mistakes; they
require a genius to correct them. Of course, we do not see
our own mistakes.

> Edward Teller, *Conversations on
> the Dark Secrets of Physics* (1991)

Nature is quite as truly "red in tooth and claw," quite
as truly an unchanging machine, quite as truly a master
against whom our revolt is beginning to succeed, quite as
truly a mere collection of things to be turned to the service
of our conscious ends. It is above all, on any reasonable
ground, a thing *to study, to know about, to see thro,* and
one can readily show that the emotionally indifferent
attitude of the scientific observer is ethically a far higher
attitude than the loving interest of the poet.

> Edward L. Thorndike, "Sentimentality in Science-Teaching,"
> *Educational Review,* January 1899

There is something fascinating about science. One gets
such wholesale returns of conjecture out of such a trifling
investment of fact.

> Mark Twain, *Life on the Mississippi* (1883)

Science is a cemetery of dead ideas.

> Miguel de Unamuno, *The Tragic Sense of Life* (1913)

Science is the most intimate school of resignation and humility, for it teaches us to bow before the seemingly most insignificant of facts.

> Miguel de Unamuno, *The Tragic Sense of Life* (1913)

True science teaches, above all, to doubt and be ignorant.

> Miguel de Unamuno, *The Tragic Sense of Life* (1913)

Science is the organized attempt of mankind to discover how things work as causal systems. The scientific attitude of mind is an interest in such questions. It can be contrasted with other attitudes, which have different interests; for instance the magical, which attempts to make things work not as material systems but as immaterial forces which can be controlled by spells; or the religious, which is interested in the world as revealing the nature of God.

> Conrad Hal Waddington, *The Scientific Attitude* (1941)

What we must vigorously oppose is the view that one may be "scientifically" contented with the conventional self-evidentness of very widely accepted value-judgments. The specific function of science . . . is just the opposite: namely, to ask questions about these things which convention makes self-evident.

> Max Weber, "The Meaning of 'Ethical Neutrality' in Sociology and Economics," *Max Weber on the Methodology of the Social Sciences* (1949)

Man is the interpreter of nature, science the right interpretation.

William Whewell, *Philosophy of the Inductive Sciences* (1840)

The aims of scientific thought are to see the general in the particular and the eternal in the transitory.

Alfred North Whitehead, in *A Dictionary of Scientific Quotations* by Alan L. Mackay (1991)

Scientific discovery consists in the interpretation for our own convenience of a system of existence which has been made with no eye to our convenience at all.

Norbert Wiener, *The Human Use of Human Beings* (1950)

"

4

Where Did We Come From and Where Are We Headed?

Man has mounted science and is now run away with. I firmly believe that before many centuries more, science will be the master of man. The engines he will have invented will be beyond his strength to control. Some day science shall have the existence of mankind in its power, and the human race commit suicide by blowing up the world.

Henry Adams, in a letter to his brother, April 11, 1862

The future of Thought, and therefore of History, lies in the hands of the physicists, and . . . the future historian must seek his education in the world of mathematical physics. A new generation must be brought up to think by new methods, and if our historical departments in the Universities cannot enter this next phase, the physical departments will have to assume this task alone.

Henry Adams, The Degradation of the Democratic Dogma (1919)

In the next twenty centuries . . . humanity may begin to understand its most baffling mystery—where are we going? The earth is, in fact, traveling many thousands of miles per hour in the direction of the constellation Hercules—to some unknown destination in the cosmos. Man must understand his universe in order to understand his destiny.

Mystery, however, is a very necessary ingredient in our lives.

Mystery creates wonder and wonder is the basis for man's desire to understand. Who knows what mysteries will be solved in our lifetime, and what new riddles will become the challenge of the new generation? Science has not mastered prophesy. We predict too much for the next year yet far too little for the next ten. Responding to challenges is one of democracy's great strengths. Our successes in space can be used in the next decade in the solution of many of our planet's problems.

Neil Armstrong, in a speech to Congress,
September 16, 1969

Reaching the Moon by three-man vessels in one long bound from Earth is like casting a thin thread across space. The main effort, in the coming decades, will be to strengthen this thread; to make it a cord, a cable, and, finally, a broad highway.

Isaac Asimov, "The Coming Decades in Space,"
The Beginning and the End (1977)

It is change, continuing change, inevitable change, that is the dominant factor in society today. No sensible decision can be made any longer without taking into account not only the world as it is, but the world as it will be.

> Isaac Asimov, "My Own View," in *The Encyclopedia of Science Fiction,* edited by Robert Holdstock (1978)

The whole of the developments and operations of analysis are now capable of being executed by machinery. . . . As soon as an Analytical Engine exists, it will necessarily guide the future course of science.

> Charles Babbage, *Passages from the Life of a Philosopher* (1864)

In fact, we will have to give up taking things for granted, even the apparently simple things. We have to learn to understand nature and not merely to observe it and endure what it imposes on us. Stupidity, from being an amiable individual defect, has become a social crime.

> John Desmond Bernal, *The Origin of Life* (1967)

If man evolved from monkeys and apes, why do we still have monkeys and apes?

> George Carlin

"Would you tell me, please, which way I ought to go from here?" "That depends a good deal on where you want to

get to," said the Cat. "I don't much care where . . . ," said
Alice. "Then it doesn't matter which way you go," said the
Cat. "So long as I get somewhere," Alice added as an
explanation. "Oh, you're sure to do that," said the Cat, "if
only you walk long enough."

> Lewis Carroll (Charles Lutwidge Dodgson),
> *Alice's Adventures in Wonderland* (1865)

Science is now the craft of the manipulation, substitution
and deflection of the forces of nature. What I see coming is
a gigantic slaughterhouse, an Auschwitz, in which valuable
enzymes, hormones, and so on will be extracted instead of
gold teeth.

> Erwin Chargaff, *Columbia Forum,* Summer 1969

The Dark Ages may return on the gleaming wings of Science.

> Winston Churchill, in a speech at
> Fulton, Missouri, March 5, 1946

Despite the dazzling success of modern technology and the
unprecedented power of modern military systems, they
suffer from a common and catastrophic fault. While
providing us with a bountiful supply of food, with great
industrial plants, with high-speed transportation, and with
military weapons of unprecedented power, they threaten
our very survival.

> Barry Commoner, *Science and Survival* (1966)

One never notices what has been done; one can only see what remains to be done. . . .

> Marie Curie, in a letter to her brother, March 18, 1894

Man with all his noble qualities, with sympathy which feels for the most debased, with benevolence which extends not only to other men but to the humblest living creature, with his god-like intellect which has penetrated into the movements and constitution of the solar system—with all these exalted powers—Man still bears in his bodily frame the indelible stamp of his lowly origin.

> Charles Darwin, *The Descent of Man* (1871)

Would it be too bold to imagine that in the great lengths of time, since the earth began to exist, perhaps millions of ages before the commencement of the history of mankind, would it be too bold to imagine that all the warm-blooded animals have arisen from one living filament which the Great First Cause endued with animality . . . and thus possessing the faculty of continuing to improve by its own inherent activity and of delivery down those improvements by generation to its posterity, world without end.

> Erasmus Darwin, *Zoonomia* (1794)

Many billions of years will elapse before the smallest, youngest stars complete their nuclear burning and shrink into white dwarfs. But with slow, agonizing finality perpetual night will surely fall.

Paul Davies, *The Last Three Minutes* (1994)

A universe that came from nothing in the big bang will disappear into nothing in the big crunch, its glorious few zillion years of existence not even a memory.

Paul Davies, *The Last Three Minutes* (1994)

We cannot delude ourselves that this new invention [the atomic bomb] will be better used. Yet it must be made, if it really is a physical possibility. If it is not made in America this year, it may be next year in Germany. There is no ethical problem; if the invention is not prevented by physical laws, it will certainly be carried out somewhere in the world. It is better, at any rate, that America should have six months start. But again, we must not pretend. Such an invention will never be kept secret; the physical principles are too obvious, and within a year every big laboratory on earth would have come to the same result. For a short time, perhaps, the U.S. government may have this power entrusted to it; but soon after it will be in less civilized hands.

Discovery magazine editorial, 1939

Everything that can be invented has been invented.

Charles H. Duell, Commissioner of U.S. Patent Office, urging President McKinley to abolish his office, 1899, in *Facts and Fallacies* by Chris Morgan and David Langford (1981)

Science and technology, like all original creations of the human spirit, are unpredictable. If we had a reliable way to label our toys good and bad, it would be easy to regulate technology wisely. But we can rarely see far enough ahead to know which road leads to damnation. Whoever concerns himself with big technology, either to push it forward or to stop it, is gambling in human lives.

> Freeman J. Dyson, *Disturbing the Universe* (1979)

We used to think that if we knew one, we knew two, because one and one are two. We are finding that we must learn a great deal more about "and."

> Sir Arthur Stanley Eddington, in *A Dictionary of Scientific Quotations* by Alan L. Mackay (1991)

There is not the slightest indication that [nuclear] energy will ever be obtainable. It would mean that the atom would have to be shattered at will.

> Albert Einstein, 1932, in *Hiroshima plus 20* edited by John Finney (1965)

I never worry about the future. It comes soon enough.

> Albert Einstein, aphorism, 1945–46

We must learn the difficult lesson that the future of mankind will only be tolerable when our course, in world affairs as in all other matters, is based upon justice and law rather than the threat of naked power.

> Albert Einstein, in a message for the
> Gandhi memorial service, February 11, 1948

I do not know how the Third World War will be fought, but I do know how the Fourth will: with sticks and stones.

> Albert Einstein (attributed)

Research during the last five years has demonstrated that cloning mammals (including humans) is theoretically impossible with today's technology—and with any technology realistically within sight.

> Michael A. Frohman, developmental biologist, State
> University of New York at Stony Brook, "The Limits of
> Genetic Engineering," in *Newsweek*, July 6, 1993

The danger of the past was that men became slaves. The danger of the future is that men may become robots.

> Erich Fromm, *The Sane Society* (1955)

Till now man has been up against Nature; from now on he will be up against his own nature.

> Dennis Gabor, *Inventing the Future* (1964)

[By 1940] the relativity theory will be considered a joke.

> George Francis Gilette, in *Fads and Fallacies in the
> Name of Science* by Martin Gardner (1929)

Life is a copiously branching bush, continually pruned by the grim reaper of extinction, not a ladder of predictable progress.

Stephen Jay Gould, *Wonderful Life* (1989)

In the lifetime of one person, we went from figuring out where we came from to figuring out how to get rid of ourselves.

Jack Horner, on the 80 years between Darwin's *On the Origin of Species* and the nuclear bomb, in *Time,* April 26, 1993

It is natural selection that gives direction to changes, orients chance, and slowly, progressively produces more complex structures, new organs, and new species. Novelties come from previously unseen association of old material. To create is to recombine.

François Jacob, in *Science,* June 10, 1977

Science should leave off making pronouncements: the river of knowledge has too often turned back on itself.

Sir James Jeans, in *The Penguin Dictionary of Twentieth-Century Quotations* by J. M. and M. J. Cohen (1993)

Like the fifteenth-century navigators, astronomers today are embarked on a voyage of exploration, charting unknown regions. The aim of this adventure is to bring back not gold or spices or silks but something more valuable: a map of

"

the universe that will tell of its origin, its texture, and its fate.

Robert Kirshner, in *Thursday's Universe*
by Marcia Bartusiak (1986)

Eight hundred life spans can bridge more than 50,000 years. But of these 800 people, 650 spent their lives in caves or worse; only the last 70 had any truly effective means of communicating with one another, only the last 6 ever saw a printed word or had any real means of measuring heat or cold, only the last 4 could measure time with any precision; only the last 2 used an electric motor; and the vast majority of the items that make up our material world were developed within the life span of the eight-hundredth person.

R. L. Lesher and G. J. Howick, *Assessing Technology
Transfer* (NASA Report SP-5067) (1966)

The world began without man, and it will end without him.

Claude Levi-Strauss, *Tristes Tropiques* (1955)

Those who are unwilling to invest in the future haven't earned one.

H. W. Lewis, *Technological Risk* (1990)

If we could first know where we are, and whither we are tending, we could better judge what to do, and how to do it.

Abraham Lincoln, in a speech in
Springfield, Illinois, June 16, 1858

None of Darwin's particular doctrines will necessarily endure the test of time and trial. Into the melting-pot must they go as often as any man of science deems it fitting. But Darwinism as the touch of nature that makes the whole world kin can hardly pass away.

Robert Ranulph Marett, *Anthropology* (1912)

There is no likelihood man can ever tap the power of the atom. . . . Nature has introduced a few foolproof devices into the great majority of elements that constitute the bulk of the world, and they have no energy to give up in the process of disintegration.

Dr. Robert Andrews Millikan, 1923, in *Facts and Fallacies* by Chris Morgan and David Langford (1981)

The views of space and time which I wish to lay before you have sprung from the soil of experimental physics, and therein lies their strength. They are radical. Henceforth space by itself, and time by itself, are doomed to fade away into mere shadows, and only a kind of union of the two will preserve an independent reality.

Herman Minkowski, September 1908, in *The Fourth Dimension and Non-Euclidean Geometry in Modern Art* by L. D. Henderson (1983)

The ancient covenant is in pieces; man knows at last that he is alone in the universe's unfeeling immensity, out of which he emerged only by chance. His destiny is nowhere spelled

"

out, nor is his duty. The kingdom above or the darkness below: it is time for him to choose.

Jacques Monod, *Chance and Necessity* (1970)

Our machines . . . will mature . . . into something transcending everything we know—in whom we can take pride when they refer to themselves as our descendants.

Hans Moravec, *Mind Children* (1988)

Do you believe then that the sciences would ever have arisen and become great if there had not beforehand been magicians, alchemists, astrologers and wizards, who thirsted and hungered after abscondite and forbidden powers?

Friedrich Nietzsche, *Die frohliche Wissenschaft* (1886)

There is no reason for any individual to have a computer in their home.

Ken Olson, President of Digital Equipment Corporation, in a speech to the World Future Society in Boston, Massachusetts, 1977

Accustomed, as we are, to laboratory work, to clear predictions, we see clearly what the more ignorant still have not realized; and I put into this category of ignorant certain men who are cultivated, but are completely unaware of science and its enormous potential, and who, by an aberration, unimaginable in the light of what we have already seen, think that the future will always have to be like the past and conclude that there will always be wars, poverty and slavery.

Jean Perrin, *La Science et l'Espérence* (1948)

Where a calculator on the ENIAC is equipped with 18,000 vacuum tubes and weighs 30 tons, computers in the future may have only 1,000 vacuum tubes and perhaps only weight $1^1/_2$ tons.

Popular Mechanics, March 1949

It is easy to be overawed by the visions of the new astronomy. Many among us would prefer to retreat into a comfortable cloud of unknowing. But if we are truly interested in knowing who we are, then we must be brave enough to accept what our senses and our reason tell us. We must enter into the universe of the galaxies and the light-years, even at the risk of spiritual vertigo, and know what after all must be known.

Chet Raymo, *The Soul of the Night* (1985)

The energy produced by the atom is a very poor kind of thing. Anyone who expects a source of power from the transformation of these atoms is talking moonshine.

Lord Ernest Rutherford, Professor of Experimental Physics at Cambridge University, in "Atom Powered World Absurd, Scientists Told," *The New York Herald Tribune*, September 12, 1933

"

. . . within half a century, machinery will perform all work—automata will direct them. The only tasks of the human race will be to make love, study, and be happy.

> *The United States Review,* 1853, in *The Timeline Book of Science* by George Ochoa and Melinda Corey (1995)

It is very hard to realize that this present universe has evolved from an unspeakably unfamiliar early condition, and faces a future extinction of endless cold or intolerable heat. The more the universe seems comprehensible, the more it also seems pointless.

> Steven Weinberg, *The First Three Minutes* (1977)

Progress imposes not only new possibilities for the future but new restrictions.

> Norbert Wiener, *The Human Use of Human Beings* (1950)

The whole procedure [of shooting rockets into space] . . . presents difficulties of so fundamental a nature, that we are forced to dismiss the notion as essentially impracticable, in spite of the author's insistent appeal to put aside prejudice and to recollect the supposed impossibility of heavier-than-air flight before it was actually accomplished.

> Richard van der Riet Wooley, British astronomer, in *Nature,* March 14, 1936

5

Imagination, Intuition, Curiosity, and Creativity

It is not once nor twice but times without number that the same ideas make their appearance in the world.

Aristotle, "On the Heavens," in *Manual of Greek Mathematics*
by T. L. Heath (1931)

If arithmetical skill is the measure of intelligence, then computers have been more intelligent than all human beings all along. If the ability to play chess is the measure, then there are computers now in existence that are more intelligent than any but a very few human beings. However, if insight, intuition, creativity, the ability to view a problem as a whole and guess the answer by the "feel" of the situation, is a measure of intelligence, computers are very unintelligent indeed. Nor can we see right now how this deficiency in computers can be easily remedied, since human beings cannot program a computer to be intuitive or creative for the very good reason that we do not know what we ourselves do when we exercise these qualities.

Isaac Asimov, *Machines That Think* (1983)

It would be the height of folly—and self-defeating—to think that things never heretofore done can be accomplished without means never heretofore tried.

Sir Francis Bacon, *Novum Organum* (1620)

In science one must search for ideas. If there are no ideas, there is no science. A knowledge of facts is only valuable in so far as facts conceal ideas: facts without ideas are just the sweepings of the brain and the memory.

Vissario Grigorievich Belinskii,
Collected Works, Volume Two (1948)

It is characteristic of science that the full explanations are often seized in their essence by the percipient scientist long in advance of any possible proof.

John Desmond Bernal, *The Origin of Life* (1967)

The important thing in science is not so much to obtain new facts as to discover new ways of thinking about them.

Sir William Lawrence Bragg, in *Beyond Reductionism*
by A. Koestler and J. R. Smithies (1958)

A popular cliché in philosophy says that science is pure
analysis or reductionism, like taking the rainbow to pieces;
and art is pure synthesis, putting the rainbow together. This
is not so. All imagination begins by analyzing nature.

Jacob Bronowski, *The Ascent of Man* (1973)

The shrewd guess, the fertile hypothesis, the courageous
leap to a tentative conclusion—these are the most valuable
coin of the thinker at work.

Jerome Seymour Bruner, *The Process of Education* (1960)

For those who do not think, it is best at least to rearrange
their prejudices once in a while.

Luther Burbank

It is well for people who think to change their minds
occasionally in order to keep them clean.

Luther Burbank

The object of opening the mind, as of opening the mouth, is
to shut it again on something solid.

G. K. Chesterton, as quoted in *New Scientist,* 1990

The only way of finding the limits of the possible is by
going beyond them into the impossible.

Arthur C. Clarke, *The Lost Worlds of 2001* (1972)

"

Questioning is the cutting edge of knowledge; assertion is
the dead weight behind the edge that gives it driving force.

>   Robin George Collingwood, *Speculum Mentis* (1924)

Humanity needs practical men, who get the most out of
their work, and, without forgetting the general good,
safeguard their own interests. But humanity also needs
dreamers, for whom the disinterested development of an
enterprise is so captivating that it becomes impossible for
them to devote their care to their own material profit.
   Without doubt, those dreamers do not deserve wealth,
because they do not desire it. Even so, a well-organized
society should assure to such workers the efficient means of
accomplishing their task, in a life freed from material care
and freely consecrated to research.

>   Marie Curie, quoted in *Madame Curie* by Eve Curie (1939)

Imagination, as well as reason, is necessary to perfection of
the philosophical mind. A rapidity of combination, a power
of perceiving analogies, and of comparing them by facts,
is the creative source of discovery. Discrimination and
delicacy of sensation, so important in physical research,
are other words for taste; and the love of nature is the same
passion, as the love of the magnificent, the sublime and the
beautiful.

>   Sir Humphrey Davy, *Parallels Between Art and Science* (1807)

Every great advance in science has issued from a new audacity of imagination.

> John Dewey, *The Quest for Certainty* (1929)

The man of science who cannot formulate a hypothesis is only an accountant of phenomena.

> Pierre L. Du Noüy, *The Road to Reason* (1949)

Genius is one per cent inspiration and ninety-nine per cent perspiration.

> Thomas Alva Edison, *Life* (1932)

If there is such a thing as luck, then *I* must be the most unlucky fellow in the world. I've never once made a lucky strike in all my life. When I get after something that I need, I start finding everything in the world that I *don't* need— one damn thing after another. I find ninety-nine things that I don't need, and then comes number one hundred, and that—at the very last—turns out to be just what I had been looking for.

> Thomas Alva Edison, in "Edison in His Laboratory"
> by M. A. Rosanoff, *Harper's,* September 1932

The fairest thing we can experience is the mysterious. It is the fundamental emotion which stands at the cradle of true art and true science. He who does not know it and can no longer wonder, no longer feel amazement, is as good as dead, a snuffed-out candle.

> Albert Einstein, "What I Believe," *Forum and Century,* 1930

Physical concepts are free creations of the human mind, and are not, however it may seem, uniquely determined by the external world.

> Albert Einstein, *The Evolution of Physics* (1938)

I have no special talents. I am only passionately curious.

> Albert Einstein, in a letter to Carl Seelig, March 11, 1952

All great achievements in science start from intuitive knowledge, namely, in axioms, from which deductions are then made. . . . Intuition is the necessary condition for the discovery of such axioms.

> Albert Einstein, in *Conversations with Einstein*
> by Alexander Moszkowski (1970)

To me it is enough to wonder at the secrets.

> Albert Einstein, on "Biography," 1991

Imagination is more important than knowledge. Knowledge is limited; imagination encircles the world.

> Albert Einstein (attributed)

The creative element in the mind of man . . . emerges in as mysterious a fashion as those elementary particles which leap into momentary existence in great cyclotrons, only to vanish again like infinitesimal ghosts.

Loren Eiseley, *The Night Country* (1971)

The whole question of imagination in science is often misunderstood by people in other disciplines. They try to test our imagination in the following way. They say, "Here is a picture of some people in a situation. What do you imagine will happen next?" When we say, "I can't imagine," they may think we have a weak imagination. They overlook the fact that whatever we are allowed to imagine in science must be consistent with everything else we know; that the electric fields and the waves we talk about are not just some happy thoughts which we are free to make as we wish, but ideas which must be consistent with all the laws of physics we know. We can't allow ourselves to seriously imagine things which are obviously in contradiction to the laws of nature. And so our kind of imagination is quite a difficult game. One has to have the imagination to think of something that has never been seen before, never been heard of before. At the same time the thoughts are restricted in a straightjacket, so to speak, limited by the conditions that come from our knowledge of the way nature really is. The problem of creating something which is new, but which is consistent with everything which has been seen before, is one of extreme difficulty.

Richard P. Feynman, *The Feynman Lectures on Physics,*
Volume Two (1963)

"

I am not really a man of science, not an observer, or an experimenter, and not a thinker. I am by temperament nothing but a conquistador . . . with the curiosity, the boldness and the tenacity that belong to that type of person.

Sigmund Freud, in *The Life and Work of Sigmund Freud*
by E. Jones (1953)

The capacity to be puzzled is . . . the premise of all creation, be it in art or in science.

Erich Fromm, "The Creative Attitude," in *Creativity and Its Cultivation* edited by Harold H. Anderson (1959)

Facts do not "speak for themselves"; they are read in the light of a theory. Creative thought, in science as much as in the arts, is the motor of changing opinion.

Stephen Jay Gould, *Perspectives in Biological Medicine* (1985)

The common perception of science as a rational activity, in which one confronts the evidence of fact with an open mind, could not be more false. Facts assume significance only within a pre-existing intellectual structure, which may be based as much on intuition and prejudice as on reason.

Walter Gratzer, in *The Guardian,* September 28, 1989

A moment's insight is sometimes worth a life's experience.

Oliver Wendell Holmes, Sr., "Iris, Her Book,"
The Professor at the Breakfast-Table (1860)

Great scientific discoveries have been made by men seeking to verify quite erroneous theories about the nature of things.

> Aldous Huxley, "Wordsworth in the Tropics,"
> in *Fifty Famous Essays* (1964)

Research is wonder.

> Steve Jones, when asked if research destroys wonder,
> in the *London Daily Telegraph*, October 26, 1982

The moment of truth, the sudden emergence of new insight, is an act of intuition. Such intuitions give the appearance of miraculous flashes, or short circuits of reasoning. In fact they may be likened to an immersed chain, of which only the beginning and the end are visible above the surface of consciousness. The diver vanishes at one end of the chain and comes up at the other end, guided by invisible links.

> Arthur Koestler, *The Act of Creation* (1964)

Nothing puzzles me more than time and space; and yet nothing troubles me less, as I never think about them.

> Charles Lamb, in a letter to Thomas Manning, January 2, 1806

How can we have any new ideas or fresh outlooks when 90 per cent of the scientists who have ever lived have still not died?

> Alan L. Mackay, in *Scientific World*, 1969

"

Creativity is what cannot wait, cannot stop, cannot backstep: faster or slower, it always goes ahead—through, alongside, above, regardless of crises or systems.

> José Roderigues Migueis, in a speech at
> Brown University, March 11, 1979

For God's sake, stop researching for a while and begin to think.

> Walter Hamilton Moberley, *The Crisis in the University* (1949)

Physics is imagination in a straight jacket.

> John Moffat, in the University of Toronto *Bulletin*,
> May 5, 1986

Man is but a reed, the most feeble thing in nature; but he is a thinking reed.

> Blaise Pascal, *Pensées* (1670)

It is his intuition, his mystical insight into the nature of things, rather than his reasoning which makes a great scientist.

> Sir Karl Popper, *The Open Society and Its Enemies* (1945)

There is another form of temptation even more fraught with danger. This is the disease of curiosity. . . . It is this which drives us on to try to discover the secrets of nature, those secrets which are beyond our understanding, which

can avail us nothing and which men should not wish
to learn.

Saint Augustine, in *Dragons of Eden* by Carl Sagan (1977)

The true scientist never loses the faculty of amazement. It is
the essence of his being.

Hans Selye, in *Newsweek,* March 31, 1958

Wonder . . . and not any expectation of advantage from its
discoveries, is the first principle which prompts mankind to
the study of Philosophy, of that science which pretends to
lay open the concealed connections that unite the various
appearances of nature.

Adam Smith, "The History of Astronomy,"
Essays on Philosophical Subjects (1795)

There are no foolish questions, and no man becomes a fool
until he has stopped asking questions.

Charles Proteus Steinmetz

We are at our human finest, dancing with our minds, when
there are more choices than two. Sometimes there are ten,
even twenty different ways to go, all but one bound to be
wrong, and the richness of selection in such situations can
lift us onto totally new ground.

Lewis Thomas, *The Medusa and the Snail:
More Notes of a Biology Watcher* (1979)

The secret of science is to ask the right question, and it is the choice of problem more than anything else that marks the man of genius in the scientific world.

Sir Henry Tizard, in *A Postscript to Science and Government* by C. P. Snow (1962)

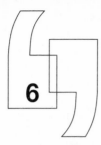

6

How Scientists Do Science

Common sense . . . has the very curious property of being more correct retrospectively than prospectively. It seems to me that one of the principal criteria to be applied to successful science is that its results are almost always obvious retrospectively; unfortunately, they seldom are prospectively. Common sense provides a kind of ultimate validation after science has completed its work; it seldom anticipates what science is going to discover.

Russell Lincoln Ackoff, *Decision Making in National Science Policy* (1968)

Basic research is like shooting an arrow into the air and, where it lands, painting a target.

Homer Adkins, in *Nature*, 1984

Every great scientific truth goes through three stages. First, people say it conflicts with the Bible. Next they say it had been discovered before. Lastly they say they always believed it.

Louis Agassiz, in A Dictionary of Scientific Quotations
by Alan L. Mackay (1991)

There is no democracy in physics. We can't say that some second-rate guy has as much right to opinion as Fermi.

Luis Walter Alvarez, in The Politics of Pure Science
by D. S. Greenberg (1967)

I have yet to see any problem, however complicated, which, when you looked at it in the right way, did not become still more complicated.

Poul Anderson, in New Scientist, September 25, 1969

Enough research will tend to support your theory.

Anonymous, in Murphy's Law by Arthur Bloch (1977),
and titled "Murphy's Law of Research"

A scientific observation is always a committed observation. It confirms or denies one's preconceptions, one's first ideas, one's plan of observation. It shows by demonstration. It structures the phenomenon. It transcends what is close at hand. It reconstructs the real after having reconstructed its representation.

Gaston Bachélard, The New Scientific Spirit (1934)

Books must follow sciences, and not sciences books.

Sir Francis Bacon, *Verulamiana* (1603)

If a man will begin with certainties, he shall end in doubts;
but if he will be content to begin with doubts, he shall end
in certainties.

Sir Francis Bacon, *The Advancement of Learning* (1605)

The men of experiment are like the ant, they only collect
and use; the reasoners resemble spiders, who make cobwebs
out of their own substance. But the bee takes the middle
course: it gathers its materials from the flowers of the garden
and field, but transforms and digests it by a power of its
own. Not unlike this is the true business of philosophy
[science]; for it neither relies solely or chiefly on the powers
of the mind, nor does it take the matter which it gathers
from natural history and mechanical experiments and lay
up in the memory whole, as it finds it, but lays it up in the
understanding altered and digested. Therefore, from a
closer and purer league between these two faculties, the
experimental and the rational (such as has never been
made), much may be hoped.

Sir Francis Bacon, *Novum Organum* (1620)

Human understanding is of its own nature prone to
abstractions and gives a substance and reality to things
which are fleeting. But to resolve nature into abstractions is

"

less to our purpose than to dissect her into parts. . . . Matter rather than forms should be the object of our attention, its configurations and changes of configuration, and simple action, and laws of action or motion; for forms are figments of the human mind, unless you will call these laws of action forms.

Sir Francis Bacon, *Novum Organum* (1620)

This science [experimental science] alone, therefore, knows how to test perfectly what can be done by nature, what by the effort of art, what by trickery, what the incantations, conjurations, invocations, deprecations, sacrifices that belong to magic mean and dream of, and what is in them, so that all falsity may be removed and the truth alone of art and nature may be retained. This science alone teaches us how to view the mad acts of magicians, that they may be not ratified but shunned, just as logic considers sophistical reasoning.

Roger Bacon, *Opus Majus* (1267)

Whenever we pride ourselves upon finding a newer, stricter way of thought or exposition; whenever we start insisting too hard upon "operationalism" or symbolic logic or any other of these very essential systems of tramlines, we lose something of the ability to think new thoughts. And equally, of course, whenever we rebel against the sterile rigidity of formal thought and exposition and let our ideas run wild, we likewise lose. As I see it, the advances in scientific thought come from a combination of loose and strict thinking, and this combination is the most precious tool of science.

Gregory Bateson, "Culture Contact and Schismogenesis" (1935), in *Steps to an Ecology of Mind: Collected Essays in Anthropology, Psychiatry, Evolution, and Epistemology* (1972)

Statistics are the triumph of the quantitative method, and the quantitative method is the victory of sterility and death.

Hilaire Belloc, *The Silence of the Sea* (1940)

There is no substitute for honest, thorough, scientific effort to get correct data (no matter how much of it clashes with preconceived ideas). There is no substitute for actually reaching a correct claim of reasoning. Poor data and good reasoning give poor results. Good data and poor reasoning give poor results. Poor data and poor reasoning give rotten results.

Edmund C. Berkeley, "Right Answers—A Short Guide for Obtaining Them," *Computers and Automation*, September 1969

In England, more than in any other country, science is felt rather than thought. . . . A defect of the English is their almost complete lack of systematic thinking. Science to them consists of a number of successful raids into the unknown.

John Desmond Bernal, *The Social Function of Science* (1938)

Our ideas are only intellectual instruments which we use to break into phenomena; we must change them when they have served their purpose, as we change a blunt lancet that we have used long enough.

> Claude Bernard, *An Introduction to the Study of Experimental Medicine* (1865)

Science allows no exceptions; without this there would be no determinism in science or rather, there would be no science at all.

> Claude Bernard, *Leçons de Pathologie Expérimentale* (1871)

How wonderful that we have met with a paradox. Now we have some hope of making progress.

> Niels Bohr, in *The Quantum Dice* by L. I. Ponomarev (1993)

Progress in science depends on new techniques, new discoveries and new ideas, probably in that order.

> Sydney Brenner, in *Nature,* May 5, 1980

. . . the fact has always been for the physicist the one ultimate thing from which there is no appeal, and in the face of which the only possible attitude is a humility almost religious.

> P. W. Bridgman, *The Logic of Modern Physics* (1927)

By the worldly standards of public life, all scholars in their work are of course oddly virtuous. They do not make wild

claims, they do not cheat, they do not try to persuade at
any cost, they appeal neither to prejudice nor to authority,
they are often frank about their ignorance, their disputes
are fairly decorous, they do not confuse what is being
argued with race, politics, sex or age, they listen patiently
to the young and to the old who both know everything.
These are the general virtues of scholarship, and they are
peculiarly the virtues of science.

> Jacob Bronowski, *Science and Human Values* (1956)

That is the essence of science: ask an impertinent question,
and you are on the way to the pertinent answer.

> Jacob Bronowski, *The Ascent of Man* (1973)

Knowledge for the sake of understanding, not merely to
prevail, that is the essence of our being. None can define its
limits, or set its ultimate boundaries.

> Vannevar Bush, *Science Is Not Enough* (1967)

A committee is a cul-de-sac down which ideas are lured and
then quietly strangled.

> Sir Barnett Cocks, in *New Scientist*, 1973

Science advances, not by the accumulation of new facts . . .
but by the continuous development of new concepts.

> James Bryant Conant, in *A Dictionary of Scientific Quotations*
> by Alan L. Mackay (1991)

"

Experiment is the interpreter of nature. Experiments never deceive. It is our judgment which sometimes deceives itself because it expects results which experiment refuses. We must consult experiment, varying the circumstances, until we have deduced general rules, for experiment alone can furnish reliable rules.

Leonardo da Vinci, in *Introductory College Physics* by Oswald Blackwood (1939)

False facts are highly injurious to the progress of science, for they often long endure; but false views, if supported by some evidence, do little harm, as every one takes a salutary pleasure in proving their falseness; and when this is done, one path towards error is closed and the road to truth is often at the same time opened.

Charles Darwin, *The Descent of Man* (1871)

My mind seems to have become a kind of machine for grinding general laws out of large collections of facts.

Charles Darwin, *Autobiography* (1876)

In October 1838 . . . I happened to read for amusement "Malthus on Population," and being well prepared to appreciate the struggle for existence which everywhere goes on from long-continued observation of the habits of animals and plants, it at once struck me that under these circumstances favorable variations would tend to be preserved, and unfavorable ones to be destroyed. The result of this would be the formation of new species. Here then I had at last got a theory by which to work.

Charles Darwin, *Autobiography* (1876)

I love fool's experiments. I am always making them.

Charles Darwin, in *The Life and Letters of Charles Darwin* edited by Sir Francis Darwin (1887)

I had . . . during many years, followed a golden rule, namely, that whenever a published fact, a new observation or thought came across me, which was opposed by my general results, to make a memorandum of it without fail and at once; for I had found by experience that such facts and thoughts were far more apt to escape from memory than favorable ones.

Charles Darwin, *The Autobiography of Charles Darwin* with original omissions restored, edited by Nora Barlow (1958)

While the artist's communication is linked forever with its original form, that of the scientist is modified, amplified, fused with the ideas and results of others.

Max Delbrück, in *The Eighth Day of Creation* by Horace Freeland Judson (1979)

I thought the following four (rules) would be enough, provided that I made a firm and constant resolution not to fail even once in the observance of them.

"

The first was never to accept anything as true if I had not evident knowledge of its being so; that is, carefully to avoid precipitancy and prejudice, and to embrace in my judgment only what presented itself to my mind so clearly and distinctly that I had no occasion to doubt it. The second, to divide each problem I examined into as many parts as was feasible, and as was requisite for its better solution. The third, to direct my thoughts in an orderly way; beginning with the simplest objects, those most apt to be known, and ascending little by little, in steps as it were, to the knowledge of the most complex; and establishing an order in thought even when the objects had no natural priority one to another. And the last, to make throughout such complete enumerations and such general surveys that I might be sure of leaving nothing out.

These long chains of perfectly simple and easy reasonings by means of which geometers are accustomed to carry out their most difficult demonstrations had led me to fancy that everything that can fall under human knowledge forms a similar sequence; and that so long as we avoid accepting as true what is not so, and always preserve the right order of deduction of one thing from another, there can be nothing too remote to be reached in the end, or too well hidden to be discovered.

René Descartes, *Discourse on Method* (1637)

It is well to know something of the manners of various peoples, in order more sanely to judge our own, and that we do not think that everything against our modes is ridiculous, and against reason, as those who have seen nothing are accustomed to think.

René Descartes, *Discourse on Method* (1637)

We have three principal means: observation of nature, reflection, and experiment. Observation gathers the facts, reflection combines them, experiment verifies the result of the combination. It is essential that the observation of nature be assiduous, that reflection be profound, and that experimentation be exact. Rarely does one see these abilities in combination. And so, creative geniuses are not common.

Denis Diderot, *On the Interpretation of Nature* (1753)

Basic research is not the same as development. A crash program for the latter may be successful; but for the former it is like trying to make nine women pregnant at once in the hope of getting a baby in a month's time.

Sir William Richard Shaboe Doll, in *New Scientist*, November 18, 1976

When you have eliminated the impossible, whatever remains, however improbable, must be the truth.

Sir Arthur Conan Doyle, *The Sign of the Four* (1890)

It is a capital mistake to theorize before one has data.
Insensibly one begins to twist facts to suit theories, instead
of theories to suit facts.

> Sir Arthur Conan Doyle, "A Scandal in Bohemia,"
> *Adventures of Sherlock Holmes* (1891)

"It is of the highest importance in the art of detection to
be able to recognize out of a number of facts which are
incidental and which are vital. . . . I would call your
attention to the curious incident of the dog in the
night-time." "The dog did nothing in the night-time."
"That was the curious incident."

> Sir Arthur Conan Doyle, "Silver Blaze,"
> *Memoirs of Sherlock Holmes* (1894)

The watchmaker to whom one gives a watch that does not
run will take it all apart and will examine each of the pieces
until he finds out which one is damaged. The physician to
whom one presents a patient cannot dissect him to establish
the diagnosis. The physicist resembles a doctor, not a
watchmaker.

> Pierre Duhem, *Revue des Questions Scientifiques*, 1897

For the truth of the conclusions of physical science,
observation is the supreme Court of Appeal. It does not
follow that every item which we confidently accept as
physical knowledge has actually been certified by the Court;

our confidence is that it would be certified by the Court if it were submitted. But it does follow that every item of physical knowledge is of a form which might be submitted to the Court. It must be such that we can specify (although it may be impracticable to carry out) an observational procedure which would decide whether it is true or not. Clearly a statement cannot be tested by observation unless it is an assertion about the results of observation. *Every item of physical knowledge must therefore be an assertion of what has been or would be the result of carrying out a specified observational procedure.*

Sir Arthur Stanley Eddington,
The Philosophy of Physical Science (1958)

Let us suppose that an ichthyologist is exploring the life of the ocean. He casts a net into the water and brings up a fishy assortment. Surveying his catch, he proceeds in the usual manner of a scientist to systematize what it reveals. He arrives at two generalizations:

(1) No sea-creature is less than two inches long.

(2) All sea-creatures have gills.

These are both true of his catch, and he assumes tentatively that they will remain true however often he repeats it.

In applying this analogy, the catch stands for the body of knowledge which constitutes physical science, and the net for obtaining it. The casting of the net corresponds to

"

observation; for knowledge which has not been or could not be obtained by observation is not admitted into physical science.

An onlooker may object that the first generalization is wrong. "There are plenty of sea-creatures under two inches long, only your net is not adapted to catch them." The ichthyologist dismisses this objection contemptuously. "Anything uncatchable by my net is *ipso facto* outside the scope of ichthyological knowledge. In short, what my net can't catch isn't fish." Or—to translate the analogy—"If you are not simply guessing, you are claiming a knowledge of the physical universe discovered in some other way than by the methods of physical science, and admittedly unverifiable by such methods. You are a metaphysician. Bah!"

Sir Arthur Stanley Eddington,
The Philosophy of Physical Science (1958)

M. A. Rosanoff: Mr. Edison, please tell me what laboratory rules you want me to observe.

Edison: Hell! There *ain't* no rules around here! We're trying to accomplish somep'n.

Thomas Alva Edison, in "Edison in His Laboratory"
by M. A. Rosanoff, Harper's, September 1932

A theory is the more impressive the greater the simplicity of its premises, the more different kinds of things it relates, and the more extended its area of applicability.

Albert Einstein, Autobiographical Notes (1979)

The scientific theorist is not to be envied. For Nature, or more precisely experiment, is an inexorable and not very friendly judge of his work. It never says "Yes" to a theory. In the most favorable cases it says "Maybe," and in the great majority of cases simply "No." If an experiment agrees with a theory it means for the latter "Maybe," and if it does not agree it means "No." Probably every theory will someday experience its "No"—most theories, soon after conception.

Albert Einstein, in *Albert Einstein: The Human Side*
by Helen Dukas and Banesh Hoffmann (1979)

In the end, science as we know it has two basic types of practitioners. One is the educated man who still has a controlled sense of wonder before the universal mystery, whether it hides in a snail's eye or within the light that impinges on that delicate organ. The second kind of observer is the extreme reductionist who is so busy stripping things apart that the tremendous mystery has been reduced to a trifle, to intangibles not worth troubling one's head about.

Loren Eiseley, "Science and the Sense of the Holy,"
The Star Thrower (1978)

It is the great beauty of our science that advancement in it, whether in a degree great or small, instead of exhausting the subject of research, opens the doors to further and more abundant knowledge, overflowing with beauty and utility.

Michael Faraday, in *A Dictionary of Scientific Quotations* by Alan L. Mackay (1991)

The world little knows how many of the thoughts and theories which have passed through the mind of a scientific investigator have been crushed in silence and secrecy by his own severe criticism and adverse examination; that in the most successful instances not a tenth of the suggestions, the hopes, the wishes, the preliminary conclusions have been realized.

Michael Faraday, in *A Dictionary of Scientific Quotations* by Alan L. Mackay (1991)

We have a habit of writing articles in scientific journals to make the work as finished as possible, to cover up all the tracks, to not worry about the blind alleys or describe how you had the wrong idea first, and so on. So there isn't any place to publish, in a dignified manner, what you actually did in order to get to do the work.

Richard P. Feynman, Nobel Lecture, 1966

No aphorism is more frequently repeated . . . than that we must ask Nature few questions, or ideally, one question at a time. The writer is convinced that this view is wholly mistaken. Nature, he suggests, will best respond to a

"

logically and carefully thought out questionnaire; indeed if we ask her a single question, she will often refuse to answer until some other topic has been discussed.

> Sir Ronald Aylmer Fisher, in *Perspectives in Medicine and Biology*, Winter 1973

It is the lone worker who makes the first advance in a subject: the details may be worked out by a team, but the prime idea is due to the enterprise, thought and perception of an individual.

> Sir Alexander Fleming, in a speech at Edinburgh University, 1951

I have had my results for a long time: but I do not yet know how I am to arrive at them.

> Karl Friedrich Gauss, in *The Mind and the Eye* by A. Arber (1954)

The road to the general, to the revelatory simplicities of science, lies through a concern with the particular, the circumstantial, the concrete, but a concern organized and directed in terms of . . . theoretical analysis . . . analyses of physical evolution, of the functioning of the nervous system, of social organization, of psychological process, of cultural patterning, and so on—and, most especially, in

"

terms of the interplay among them. That is to say, the road lies, like any genuine Quest, through a terrifying complexity.

Clifford Geertz, "The Impact of the Concept of Culture on the Concept of Man" (1966), *The Interpretation of Cultures* (1973)

In 1963, when I assigned the name "quark" to the fundamental constituents of the nucleon, I had the sound first, without the spelling, which could have been "kwork." Then, in one of my occasional perusals of *Finnegans Wake,* by James Joyce, I came across the word "quark" in the phrase "Three quarks for Muster Mark." Since "quark" (meaning, for one thing, the cry of a gull) was clearly intended to rhyme with "Mark," as well as "bark" and other such words, I had to find an excuse to pronounce it as "kwork." But the book represents the dreams of a publican named Humphrey Chimpden Earwicker. Words in the text are typically drawn from several sources at once, like the "portmanteau words" in *Through the Looking Glass.* From time to time, phrases occur in the book that are partially determined by calls for drinks at the bar. I argued, therefore, that perhaps one of the multiple sources of the cry "Three quarks for Muster Mark" might be "Three quarts for Mister Mark," in which case the pronunciation "kwork" would not be totally unjustified. In any case, the number three fitted perfectly the way quarks occur in nature.

Murray Gell-Mann, *The Quark and the Jaguar* (1994)

As soon as any one belongs to a narrow creed in science, every unprejudiced and true perception is gone.

> Johann Wolfgang von Goethe, May 18, 1824,
> in *Conversations with Goethe* by Johann Peter Eckermann

Hypotheses are the scaffolds which are erected in front of a building and removed when the building is completed. They are indispensable to the worker; but he must not mistake the scaffolding for the building.

> Johann Wolfgang von Goethe, *Maxims and Reflections* (1893)

When an inquiry becomes so convoluted, we must suspect that we are proceeding in the wrong way. We must return to go, change gears, and reformulate the problem, not pursue every new iota of information or nuance of argument in the old style, hoping all the time that our elusive solution simply awaits a crucial item, yet undiscovered.

> Stephen Jay Gould, *The Flamingo's Smile* (1985)

The way to do research is to attack the facts at the point of greatest astonishment.

> Celia Green, *The Decline and Fall of Science* (1972)

When an apparent fact runs contrary to logic and common sense, it's obvious that you have failed to interpret the fact correctly.

> Robert A. Heinlein, *Orphans of the Sky* (1963)

It is often the scientist's experience that he senses the nearness of truth when . . . connections are envisioned. A connection is a step toward simplification, unification. Simplicity is indeed often the sign of truth and a criterion of beauty.

Mahlon Hoagland, *Toward the Habit of Truth* (1990)

The truth is, the science of Nature has been already too long made only a work of the brain and the fancy: It is now high time that it should return to the plainness and soundness of observation on material and obvious things.

Robert Hooke, *Micrographia* (1665)

Astronomers always work in the past; because light takes time to move from one place to another, they see things as they were, not as they are.

Neale E. Howard, *The Telescope Handbook
and Star Atlas* (1967)

The astronomer is severely handicapped as compared with other scientists. He is forced into a comparatively passive role. He cannot invent his own experiments as the physicist, the chemist or the biologist can. He cannot travel about the Universe examining the items that interest him. He cannot, for example, skin a star like an onion and see how it works inside.

Fred Hoyle, *The Nature of the Universe* (1950)

I don't see the logic of rejecting data just because they seem incredible.

> Fred Hoyle, in *Astronomy Transformed*
> by D. O. Edge and M. J. Mulkay (1976)

To speculate without facts is to attempt to enter a house of which one has not the key, by wandering aimlessly round and round, searching the walls and now and then peeping through the windows. Facts are the key.

> Julian Huxley, "Heredity," *Essays in Popular Science* (1923)

Science is nothing but trained and organized common sense, differing from the latter only as a veteran may differ from a raw recruit: and its methods differ from those of common sense only as far as the guardsman's cut and thrust differ from the manner in which a savage wields his club.

> T. H. Huxley, "The Method of Zadig,"
> *Collected Essays* (1893–94)

An article in *Bioscience* in November 1987 by Julie Ann Miller claimed the cortex was a "quarter-meter square." That is napkin-sized, about ten inches by ten inches. *Scientific American* magazine in September 1992 upped the ante considerably with an estimate of $1\frac{1}{2}$ square meters; that's a square of brain forty inches on each side, getting close to the card-table estimate. A psychologist at the University of Toronto figured it would cover the floor of his

"

living room (I haven't seen his living room), but the prize winning estimate so far is from the British magazine *New Scientist's* poster of the brain published in 1993 which claimed that the cerebral cortex, if flattened out, would cover a tennis court. How can there be such disagreement? How can so many experts not know how big the cortex is? I don't know, but I'm on the hunt for an expert who will say the cortex, when fully spread out, will cover a football field. A Canadian football field.

Jay Ingram, *The Burning House:*
Unlocking the Mysteries of the Brain (1994)

Science, like life, feeds on its own decay. New facts burst old rules; then newly divined conceptions bind old and new together into a reconciling law.

William James, *The Will to Believe and*
Other Essays in Popular Philosophy (1910)

It is like using Scotch tape to pull together a mule, a whale, a tiger and a giraffe.

Michio Kaku, on scientific research,
in *U.S. News & World Report*, May 9, 1994

There are many examples of old, incorrect theories that stubbornly persisted, sustained only by the prestige of foolish but well-connected scientists. . . . Many of these theories have been killed off only when some decisive experiment exposed their incorrectness. . . . Thus the yeoman work of any science, and especially physics, is done by the experimentalist, who must keep the theoreticians honest.

Michio Kaku, *Hyperspace* (1995)

The purpose of models is not to fit the data but to sharpen the questions.

Samuel Karlin, 11th R. A. Fisher Memorial Lecture,
Royal Society, April 20, 1983

Without the hard little bits of marble which are called "facts" or "data" one cannot compose a mosaic; what matters, however, are not so much the individual bits, but the successive patterns into which you arrange them, then break them up and rearrange them.

Arthur Koestler, *The Act of Creation* (1964)

It would be interesting to inquire how many times essential advances in science have first been made possible by the fact that the boundaries of special disciplines were not respected. . . . Trespassing is one of the most successful techniques in science.

Wolfgang Köhler, *Dynamics in Psychology* (1940)

The scientific mind does not so much provide the right answers as ask the right questions.

Claude Levi-Strauss, *The Raw and the Cooked* (1964)

It is a good morning exercise for a research scientist to discard a pet hypothesis every day before breakfast. It keeps him young.

> Konrad Lorenz, *On Aggression* (1966)

. . . the world of science may be the only existing participatory democracy.

> S. E. Luria, *A Slot Machine, A Broken Test Tube:*
> *An Autobiography* (1984)

Everyone knows that in research there are no final answers, only insights that allow one to formulate new questions.

> S. E. Luria, *A Slot Machine, A Broken Test Tube:*
> *An Autobiography* (1984)

If a scientist uncovers a publishable fact, it will become central to his theory.

> "Mann's Law," in "Advanced Researchmanship,"
> *Murphy's Law Book Two* by Arthur Bloch (1980)

The difference between a good observer and one who is not good is that the former is quick to take a hint from the facts, from his early efforts to develop skill in handling them, and quick to acknowledge the need to revise or alter the conceptual framework of his thinking. The other—the poor observer—continues dogmatically onward with his original thesis, lost in a maze of correlations, long after the

facts have shrieked in protest against the interpretation put
upon them.

Elton Mayo, *The Social Problems of an*
Industrial Civilization (1945)

In science, all facts, no matter how trivial or banal, enjoy
democratic equality.

Mary McCarthy, "The Fact in Fiction," *On the Contrary* (1961)

. . . in real life mistakes are likely to be irrevocable. Computer
simulation, however, makes it economically practical to
make mistakes on purpose. If you are astute, therefore, you
can learn much more than they cost. Furthermore, if you
are at all discreet, no one but you need ever know you
made a mistake.

John McLeod and John Osborn, in *Natural Automata and*
Useful Simulations edited by H. H. Pattee et al. (1966)

The scientist values research by the size of its contribution
to that huge, logically articulated structure of ideas which
is already, though not yet half built, the most glorious
accomplishment of mankind.

Sir Peter Medawar, *The Art of the Soluble* (1967)

The best person to decide what research shall be done is
the man who is doing the research. The next best is the
head of the department. After that you leave the field of

best persons and meet increasingly worse groups. The first of these is the research director, who is probably wrong more than half the time. Then comes a committee which is wrong most of the time. Finally there is a committee of company vice-presidents, which is wrong all the time.

> Charles Edward Kenneth Mees, 1935, in *Biographical Memoirs of Fellows of the Royal Society*, 1961

Knowledge is an attitude, a passion, actually an illicit attitude. For the compulsion to know is just like dipsomania, erotomania, homicidal mania, in producing a character that is out of balance. It is not true that the scientist goes after truth. It goes after him.

> Robert Musil, *Der Mann ohne Eigenschafter* (1930)

Knowledge is the death of research.

> Hermann Walther Nernst, in *The Dictionary of Scientific Biography* edited by C. G. Gillespie (1981)

By always thinking unto them. I keep the subject constantly before me and wait till the first dawnings open little by little into the full light.

> Sir Isaac Newton, on how he made discoveries, in *Nature*, September 4, 1965

Discovery follows discovery, each both raising and answering questions, each ending a long search, and each providing the new instruments for a new search.

J. Robert Oppenheimer, "Prospects in the Arts and Sciences," in *Fifty Famous Essays* (1964)

For every fact there is an *infinity* of hypotheses. The more you *look* the more you *see*.

Robert M. Pirsig, *Zen and the Art of Motorcycle Maintenance* (1974)

Nothing is more interesting to the true theorist than a fact which directly contradicts a theory generally accepted up to that time, for this is his particular work.

Max Planck, *A Survey of Physics* (1925)

Anybody who has been seriously engaged in scientific work of any kind realizes that over the entrance to the gates of the temple of science are written the words: *Ye must have faith*. It is a quality which the scientist cannot dispense with.

Max Planck, *Where Is Science Going?* (1932)

Experimenters are the shocktroops of science.

Max Planck, *Scientific Autobiography and Other Papers* (1949)

"

Experiment is the sole source of truth. It alone can teach us something new; it alone can give us certainty.

Henri Poincaré, *The Foundations of Science* (1921)

But I shall certainly admit a system as empirical or scientific only if it is capable of being *tested* by experience. These considerations suggest that not the *verifiability* but the *falsifiability* of a system is to be taken as a criterion of demarcation. . . . It must be possible for an empirical scientific system to be refuted by experience.

Sir Karl Popper, *The Logic of Scientific Discovery* (1959)

Theories are nets cast to catch what we call "the world": to rationalize, to explain, and to master it. We endeavor to make the mesh ever finer and finer.

Sir Karl Popper, *The Logic of Scientific Discovery* (1959)

Theories and schools, like microbes and globules, devour each other and by their struggle ensure the continuing of life.

Marcel Proust, *Cities of the Plain* (1927)

It is the tension between the scientist's laws and his own attempted breaches of them that powers the engines of science and makes it forge ahead.

W. V. O. Quine, *Quiddities* (1987)

The whole history of physics proves that a new discovery is quite likely lurking at the next decimal place.

F. K. Richtmeyer, in *Science,* 1932

We are peeling an onion layer by layer, each layer uncovering in a sense another universe, unexpected, complicated, and—as we understand more—strangely beautiful.

M. A. Ruderman and A. H. Rosenfeld,
"An Explanatory Statement on Elementary Particle Physics,"
American Scientist, June 1960

When a man of science speaks of his "data," he knows very well in practice what he means. Certain experiments have been conducted, and have yielded certain observed results, which have been recorded. But when we try to define a "datum" theoretically, the task is not altogether easy. A datum, obviously, must be a fact known by perception. But it is very difficult to arrive at a fact in which there is no element of inference, and yet it would seem improper to call something a "datum" if it involved inferences as well as observation. This constitutes a problem. . . .

Bertrand Russell, *The Analysis of Matter* (1954)

Profound thoughts arise only in debate, with a possibility of counterargument, only when there is a possibility of expressing not only correct ideas but also dubious ideas.

Andrei Dmitrievich Sakharov, *Progress, Coexistence,
and Intellectual Freedom* (1968)

"

You must remember that nothing happens quite by chance. It's a question of accretion of information and experience . . . it's just chance that I happened to be here at this particular time when there was available and at my disposal the great experience of all the investigators who plodded along for a number of years.

> Jonas Salk, on his discovery of the polio vaccine, in
> *Breakthrough: The Saga of Jonas Salk* by Richard Carter (1965)

When all beliefs are challenged together, the just and necessary ones have a chance to step forward and to re-establish themselves alone.

> George Santayana, *The Life of Reason:*
> *Reasons in Science* (1905–06)

. . . the task is . . . not so much to see what no one has yet seen; but to think what nobody has yet thought, about that which everybody sees.

> Erwin Schrödinger, in *Problems of Life*
> by L. Bertalanffy (1952) [See Albert Szent-Györgyi]

Discovery consists of seeing what everybody has seen and thinking what nobody has thought.

> Albert Szent-Györgyi, in *The Scientist Speculates*
> edited by John Good (1962) [See Erwin Schrödinger]

. . . research in applied science leads to reforms, research in pure science leads to revolutions, and revolutions, whether political or industrial, are exceedingly profitable things if you are on the winning side.

J. J. Thomson, in *J. J. Thomson* by Lord Rayleigh (1943)

I am never content until I have constructed a mechanical model of the subject I am studying. If I succeed in making one, I understand; otherwise I do not.

William Thomson (Lord Kelvin), in *A Dictionary of Scientific Quotations* by Alan L. Mackay (1991)

The great difference between science and technology is a difference of initial attitude. The scientific man follows his method whithersoever it may take him. He seeks acquaintance with his subject-matter, and he does not at all care about what he shall find, what shall be the content of his knowledge when acquaintance-with is transformed into knowledge-about. The technologist moves in another universe; he seeks the attainment of some determinate end, which is his sole and obsessing care; and he therefore takes no heed of anything that he cannot put to use as means toward that end.

Edward B. Titchener, *Systemic Psychology* (1929)

The instinctive tendency of the scientific man is toward the existential substrate that appears when use and purpose—

"

cosmic significance, artistic value, social utility, personal preference—have been removed. He responds positively to the bare "what" of things; he responds negatively to any further demand for interest or appreciation.

Edward B. Titchener, *Systemic Psychology* (1929)

The outcome of any serious research can only be to make two questions grow where only one grew before.

Thorstein Veblen, *The Place of Science in Modern Civilization and Other Essays* (1919)

Every scientist is an agent of cultural change. He may not be a champion of change; he may even resist it, as scholars of the past resisted the new truths of historical geology, biological evolution, unitary chemistry, and non-Euclidean geometry. But to the extent that he is a true professional, the scientist is inescapably an agent of change. His tools are the instruments of change—skepticism, the challenge to establish authority, criticism, rationality, and individuality.

Alexander Vucinich, *Science in Russian Culture: A History to 1860* (1963)

These checks—war, disease, famine and the like—must, it occurred to me, act on animals as well as man. Then I thought of the enormously rapid multiplication of animals, causing these checks to be very much more effective in them than in the case of man; and while pondering vaguely on

this fact there suddenly flashed upon me the idea of the
survival of the fittest—that the individuals removed by
these checks must be on the whole inferior to those that
survived. In the two hours that elapsed before my ague fit
was over, I had thought out almost the whole of the theory:
and the same evening I sketched the draft on paper, and in
the two succeeding evenings wrote it out in full, and sent it
by the next post to Mr. Darwin.

> Alfred Russell Wallace, in *Darwin and Butler*
> by B. Willey (1960)

. . . Science moves with the spirit of an adventure characterized
both by youthful arrogance and by the belief that the truth,
once found, would be simple as well as pretty.

> James D. Watson, *The Double Helix* (1968)

Science seldom proceeds in the straightforward logical
manner imagined by outsiders. Instead, its steps forward
(and sometimes backward) are often very human events in
which personalities and cultural traditions play major roles.

> James D. Watson, *The Double Helix* (1968)

In science, each of us knows that what he has accomplished
will be antiquated in ten, twenty, fifty years. That is the fate
to which science is subjected; it is the very *meaning* of
scientific work, to which it is devoted in a quite specific
sense, as compared with other spheres of culture for which

"

in general the same holds. Every scientific "fulfillment" raises new "questions"; it *asks* to be "surpassed" and outdated. Whoever wishes to serve science has to resign himself to this fact. Scientific works certainly can last as "gratifications" because of their artistic quality, or they may remain important as a means of training. Yet they will be surpassed scientifically—let that be repeated—for it is our common fate and, more, our common goal. We cannot work without hoping that others will advance further than we have.

> Max Weber, "Science as a Vocation" (1919),
> *Max Weber: Essays in Sociology*
> edited by H. H. Gerth and C. Wright Mills (1946)

Basic research is when you don't know what you are doing.

> Charles G. Wilson, in *Nature*, 1976

Good and Evil, Life and Death

Life is pleasant. Death is peaceful. It's the transition that's troublesome.

> Isaac Asimov, "How Easy to See the Future,"
> *Natural History,* April 1975

Now we are all sons of bitches.

> Kenneth Tompkins Bainbridge, on the first atomic bomb test,
> July 16, 1945, in *The Decision to Drop the Bomb*
> by Len Giovanitti and Fred Freed (1965)
> [See Thomas Farrell and J. Robert Oppenheimer]

The world has achieved brilliance without wisdom, power without conscience. Ours is a world of nuclear giants and ethical infants.

> General Omar Bradley, in a speech in Boston, Massachusetts,
> November 10, 1948

Scientists, therefore, are responsible for their research not only intellectually but also morally . . . the results of quantum mechanics and relativity theory have opened up two very different paths for physics to pursue. They may lead us—to put it in extreme terms—to the Buddha or to the bomb, and it is up to each of us to decide which path to take.

Fritjof Capra, *The Turning Point* (1982)

As soon as questions of will or decision or reason or choice of action arise, human science is at a loss.

Noam Chomsky, in a television interview,
in *Listener,* April 6, 1978

It would be impossible, it would be against the scientific spirit. . . . Physicists should always publish their researches completely. If our discovery has a commercial future that is a circumstance from which we should not profit. If radium is to be used in the treatment of disease, it is impossible for us to take advantage of that.

Marie Curie, in a discussion with her husband, Pierre, about the patenting of radium, in *Marie Curie* by Eve Curie (1939)

Science through its physical technological consequences is now determining the relations which human beings, severally and in groups, sustain to one another. If it is incapable of developing moral techniques which will also determine those relations, the split in modern culture goes so deep that not only democracy but all civilized values are doomed.

John Dewey, *Freedom and Culture* (1939)

Natural science has outstripped moral and political science. That is too bad; but it is a fact, and the fact does not disappear because we close our eyes to it.

John Foster Dulles, *War or Peace* (1950)

Our death is not an end if we can live on in our children and the younger generation. For they are us, our bodies are only wilted leaves on the tree of life.

Albert Einstein, in a letter to Dutch physicist
Heike Kamerlingh-Omnes' widow, February 25, 1926

Morality is of the highest importance—but for us, not for God.

Albert Einstein, in a letter to a banker in Colorado,
August 1927, in *Albert Einstein, the Human Side*
by Helen Dukas and Banesh Hoffman (1979)

I have never looked upon ease and happiness as ends in themselves—such an ethical basis I call the ideal of a pigsty. . . . The ideals which have lighted my way, and time after time have given me new courage to face life cheerfully, have been Kindness, Beauty, and Truth.

Albert Einstein, "What I Believe," *Forum and Century,* 1930

Humanity has every reason to place the proclaimers of high moral standards and values above the discoverers of objective truth. What humanity owes to personalities like Buddha, Moses, and Jesus ranks for me higher than all the achievements of the inquiring constructive mind.

> Albert Einstein, September 1937, in *Albert Einstein, the Human Side* by Helen Dukas and Banesh Hoffman (1979)

Some recent work by E. Fermi and L. Szilard which has been communicated to me in manuscript leads me to expect that the element uranium may be turned into a new and important source of energy in the near future. Certain aspects of the situation which has arisen seem to call for watchfulness and, if necessary, quick action on the part of the Administration. . . . In the course of the last four months it has been made almost certain . . . that it may become possible to set up a nuclear chain reaction in a large mass of uranium, by which vast elements would be generated. . . . This new phenomenon would lead also to the construction of bombs.

> Albert Einstein, in a letter to President Franklin D. Roosevelt, 1939

We scientists, whose tragic destiny it has been to help make the methods of annihilation ever more gruesome and more effective, must consider it our solemn and transcendent duty to do all in our power to prevent these weapons from

being used for the brutal purpose for which they were
invented.

> Albert Einstein, in *The New York Times*, August 29, 1948

The most important human endeavor is the striving for
morality in our actions. Our inner balance and even our
very existence depend on it. Only morality in our actions
can give beauty and dignity to life.

> Albert Einstein, in a letter to a minister
> in Brooklyn, N.Y., November 20, 1950

Now he has departed from this strange world a little ahead
of me. That signifies nothing. For us believing physicists the
distinction between past, present, and future is only a stubbornly persistent illusion.

> Albert Einstein, in a letter of condolence to the family
> of his friend Michele Besso, March 21, 1955

Thirty seconds after the explosion came, first the air blast
pressing hard against people and things, to be followed
almost immediately by the strong, sustained awesome roar
which warned of doomsday and made us feel that we puny
things were blasphemous to dare tamper with the forces
heretofore reserved to the Almighty.

> Thomas Farrell, official report on the first atom bomb test,
> Alamogordo, New Mexico, July 16, 1945 [See Kenneth
> Tompkins Bainbridge and J. Robert Oppenheimer]

"

Only science can hope to keep technology in some sort of moral order.

> Edgar Z. Friedenberg, "The Impact of the School,"
> *The Vanishing Adolescent* (1959)

What a curious picture it is to find man, homo sapiens, of divine origin, we are told, seriously considering going underground to escape the consequences of his own folly. With a little wisdom and foresight, surely it is not yet necessary to forsake life in the fresh air and in the warmth of the sunlight. What a paradox if our cleverness in science should force us to live underground with the moles.

> Senator J. William Fulbright, in a speech to the Foreign Policy
> Association, New York City, October 20, 1945

There can be no final truth in ethics any more than in physics, until the last man has had his experience and his say.

> William James, "The Moral Philosopher and the Moral Life,"
> *The Will to Believe and Other Essays in
> Popular Philosophy* (1910)

. . . the truth is that the knowledge of external nature and of the sciences which that knowledge requires or includes, is not the great or the frequent business of the human mind. Whether we provide for action or conversation, whether we wish to be useful or pleasing, the first requisite is the religious and moral knowledge of right and wrong; the next is an acquaintance with the history of mankind, and with those examples which may be said to embody truth, and prove by events the reasonableness of opinions. Prudence and justice are virtues, and excellencies, of all times and of all places; we are perpetually moralists, but we are geometricians only by chance. Our intercourse with intellectual nature is necessary; our speculations upon matter are voluntary, and at leisure. Physical knowledge is of such rare emergence, that one man may know another half his life without being able to estimate his skill in hydrostatics or astronomy; but his moral and prudential character immediately appears.

Dr. Samuel Johnson, in *Lives of the Poets* (1779–81)

I am sorry to say that there is too much point to the wisecrack that life is extinct on other planets because their scientists were more advanced than ours.

John F. Kennedy, in a speech on December 11, 1959

The means by which we live have outdistanced the ends for which we live. Our scientific power has outrun our spiritual power. We have guided missiles and misguided men.

Martin Luther King, Jr., *Strength to Love* (1963)

Without meaning to belittle the wonders of science, I do not think they can absolve mankind of suffering, desire, madness, and death.

Lewis H. Lapham, in a speech at
Northwestern Medical School, June 1987

"

Science itself is a humanist in the sense that it doesn't discriminate between human beings, but it is also morally neutral. It is no better or worse than the ethos for which it is used.

> Max Lerner, "Manipulating Life,"
> *New York Post*, January 24, 1968

Atomic energy bears that same duality that has faced man from time immemorial, a duality expressed in the Book of Books thousands of years ago: "See, I have set before thee this day life and good and death and evil . . . therefore choose life."

> David E. Lilienthal, in *This I Do Believe*
> edited by Edward R. Murrow (1949)

Science without conscience is but death of the soul.

> Michel de Montaigne, *Essays* (1580)

It is time that science, having destroyed the religious basis for morality, accepted the obligation to provide a new and rational basis for human behavior—a code of ethics concerned with man's needs on earth, not his rewards in heaven.

> Bernard More Oliver, "Toward a New Morality,"
> *IEEE Spectrum*, 1972

We knew the world would not be the same. A few people laughed, a few people cried. Most people were silent. I remembered the line from the Hindu scripture, the Bhagavad Gita. . . . "I am become Death, the shatterer of worlds."

> J. Robert Oppenheimer, on the first atomic bomb test,
> July 16, 1945, in *The Decision to Drop the Bomb*
> by Len Giovanitti and Fred Freed (1965)
> [See Kenneth Tompkins Bainbridge and Thomas Farrell]

Despite the vision and the farseeing wisdom of our wartime heads of state, the physicists felt a peculiarly intimate responsibility for suggesting, for supporting, and in the end, in large measure, for achieving, the realization of atomic weapons. Nor can we forget that these weapons, as they were in fact used, dramatized so mercilessly the inhumanity and evil of modern war. In some sort of crude sense which no vulgarity, no humor, no overstatement can quite extinguish, the physicists have known sin; and this is a knowledge which they cannot lose.

> J. Robert Oppenheimer, "Physics in the Contemporary
> World," The Arthur Dehon Little Memorial Lecture at the
> Massachusetts Institute of Technology, November 25, 1947

When you see something that is technically sweet, you go ahead and do it and you argue about what to do about it only after you have had your technical success.

> J. Robert Oppenheimer, in *Tongues of Conscience:
> Weapons Research and the Scientist's Dilemma* (1969)

The atomic bomb . . . made the prospect of future war unendurable. It has led us up those last few steps to the mountain pass; and beyond there is a different country.

> J. Robert Oppenheimer, in *The Making of the Atomic Bomb*
> by Richard Rhodes (1987)

Science, by itself, cannot supply us with an ethic. It can show us how to achieve a given end, and it may show us that some ends cannot be achieved.

> Bertrand Russell, "The Science to Save Us from Science,"
> *The New York Times Magazine,* March 19, 1950

What is false in the science of facts may be true in the science of values.

> George Santayana, *Interpretations of Poetry and Religion* (1900)

People must understand that science is inherently neither a potential for good nor for evil. It is a potential to be harnessed by man to do his bidding.

> Glenn T. Seaborg, in an Associated Press interview,
> September 29, 1964

"

Nature is neutral. Man has wrested from nature the power to make the world a desert or to make the deserts bloom. There is no evil in the atom; only in men's souls.

> Adlai Stevenson, in a speech in Hartford, Connecticut,
> September 18, 1952

Scourges, pestilence, famine, earthquakes, and wars are to be regarded as blessings, since they serve to prune away the luxuriant growth of the human race.

> Tertullian (Quintus Septimius Florens Tertullianus),
> in *The Timeline Book of Science*
> by George Ochoa and Melinda Corey (1995)

[Science] . . . gives us no answer to our question, what shall we do and how shall we live?

> Leo Tolstoy, *What Is Art?* (1898)

Science is out of the reach of morals, for her eyes are fixed upon eternal truths. Art is out of the reach of morals, for her eyes are fixed upon things beautiful and immortal and ever-changing. To *morals* belong the lower and less intellectual spheres.

> Oscar Wilde, *The Critic as Artist* (1891)

The great scientists have been occupied with values—it is only their vulgar followers who think they are not. If scientists like Descartes, Newton, Einstein, Darwin, and Freud don't "look deeply into experience," what do they do? They have imaginations as powerful as any poet's and some of them were first-rate writers as well. How do you draw the line between *Walden* and *The Voyage of the Beagle*? The product of the scientific imagination is a new vision of relations—like that of the artistic imagination.

Edmund Wilson, in a letter to Allen Tate, July 20, 1931

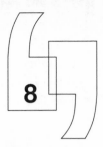

8

What Can We Know, and What Must Remain a Mystery?

Can we actually "know" the universe? My God, it's hard enough finding your way around in Chinatown.

Woody Allen, *Getting Even* (1971)

The young specialist in English Lit . . . lectured me severely on the fact that in *every* century people have thought they understood the Universe at last, and in *every* century they were proved to be wrong. It follows that the one thing we can say about our modern "knowledge" is that it is *wrong*.

. . . My answer to him was, " . . . when people thought the Earth was flat, they were wrong. When people thought the Earth was spherical they were wrong. But if you think that thinking the Earth is spherical is just as wrong as thinking the Earth is flat, then your view is wronger than both of them put together."

Isaac Asimov, *The Relativity of Wrong* (1996)

What is the *universe* but the question, what is the *universe*?

Astronomia, an exhibit at the Hayden Planetarium,
New York City, December 1987

For man being the minister and interpreter of nature, acts
and understands so far as he has observed of the order, the
works and mind of nature, and can proceed no further; for
no power is able to loose or break the chain of causes, nor
is nature to be conquered but by submission: whence those
twin intentions, human knowledge and human power, are
really coincident; and the greatest hindrance to works is the
ignorance of causes.

Sir Francis Bacon, *The Great Instauration*

Every answer given arouses new questions. The progress
of science is matched by an increase in the hidden and
mysterious.

Leo Baeck, *Judaism and Science* (1949)

One aim of the physical sciences has been to give an exact
picture of the material world. One achievement of physics
in the twentieth century has been to prove that that aim is
unattainable.

Jacob Bronowski, *The Ascent of Man* (1973)

"

The astronomers with all their hypotheses give us no satisfying or abiding conception of the Universe. We are left as bewildered as ever.

Sir James Crichton Browne,
From the Doctor's Notebook (1937)

Beyond these are other suns, giving light and life to other systems, not a thousand, or two thousand merely, but multiplied without end, and ranged all around us, at immense distances from each other, attended by ten thousand times ten thousand worlds, all in rapid motion; yet calm, regular and harmonious—all space seems to be illuminated, and every particle of light a world. . . . And yet all this vast assemblages of suns and worlds may bear no greater proportion to what lies beyond the utmost boundaries of human vision, than a drop of water to the ocean.

Elijah H. Burritt, *The Geography of the Heavens* (1863)

That knowledge is not happiness, and science
But an exchange of ignorance for that
Which is another kind of ignorance.

George Gordon, Lord Byron, *Manfred* (1817)

Science is wonderfully equipped to answer the question "How?" but gets terribly confused when you ask the question "Why?"

Erwin Chargaff, *Columbia Forum*, Summer 1969

As for a future life, every man must judge for himself between conflicting vague probabilities.

> Charles Darwin, in *The Life and Letters of Charles Darwin*
> edited by Sir Francis Darwin (1887)

The universe is like a safe to which there is a combination. But the combination is locked up in the safe.

> Peter De Vries, *Let Me Count the Ways* (1965)

The eternal mystery of the world is its comprehensibility. . . . The fact that it is comprehensible is a miracle.

> Albert Einstein, "Physics and Reality,"
> *Franklin Institute Journal,* March 1936

It has been said repeatedly that one can never, try as he will, get around to the front of the universe. Man is destined to see only its far side, to realize nature only in retreat.

> Loren Eiseley, "The Innocent Fox," *The Star Thrower* (1978)

All is riddle, and the key to a riddle is another riddle.

> Ralph Waldo Emerson, "Illusions," *The Conduct of Life* (1860)

In place of science, the Eskimo has only magic to bridge the gap between what he can understand and what is not known. Without magic, his life would be one long panic.

> Peter Farb, *Man's Rise to Civilization* (1968)

In its efforts to learn as much as possible about nature, modern physics has found that certain things can never be "known" with certainty. Much of our knowledge must always remain uncertain. The most we can know is in terms of probabilities.

Richard P. Feynman, *The Feynman Lectures on Physics,*
Volume One (1963)

I often use the analogy of a chess game: one can learn all the rules of chess, but one doesn't know how to play well. . . . The present situation in physics is as if we know chess, but we don't know one or two rules. But in this part of the board where things are in operation, those one or two rules are not operating much and we can get along pretty well without understanding those rules. That's the way it is, I would say, regarding the phenomena of life, consciousness and so forth.

Richard P. Feynman, in *Superstrings: A Theory of Everything?*
by P. C. W. Davies and Julian Brown (1988)

What I am going to tell you about is what we teach our physics students in the third or fourth year of graduate school. . . . It is my task to convince you not to turn away because you don't understand it. You see my physics students don't understand it. . . . That is because I don't understand it. Nobody does.

Richard P. Feynman,
QED, The Strange Theory of Light and Matter (1990)

I can live with doubt and uncertainty. I think it's much more interesting to live not knowing than to have answers which might be wrong.

Richard P. Feynman

The history of thought should warn us against concluding that because the scientific theory of the world is the best that has yet been formulated, it is necessarily complete and final. We must remember that at bottom the generalizations of science or, in common parlance, the laws of nature are merely hypotheses devised to explain that ever-shifting phantasmagoria of thought which we dignify with the high-sounding names of the world and the universe. In the last analysis magic, religion, and science are nothing but theories of thought.

James George Frazer, *The Golden Bough* (1890)

It would be a mistake to suppose that a science consists entirely of strictly proved theses, and it would be unjust to require this. Only a disposition with a passion for authority will raise such a demand, someone with a craving to replace his religious catechism by another, though it is a scientific one. Science has only a few apodeictic propositions in its catechism: the rest are assertions promoted by it to some particular degree of probability. It is actually a sign of a scientific mode of thought to find satisfaction in these approximations to certainty and to be able to pursue

"

constructive work further in spite of the absence of final confirmation.

> Sigmund Freud, *Introductory Lectures on Psycho-Analysis* (1916–17)

. . . poets are masters of us ordinary men, in knowledge of the mind, because they drink at streams which we have not yet made accessible to science.

> Sigmund Freud, in *A Dictionary of Scientific Quotations* by Alan L. Mackay (1991)

I have no doubt that in reality the future will be vastly more surprising than anything I can imagine. Now my own suspicion is that the universe is not only queerer than we suppose, but queerer than we can suppose.

> J. B. S. Haldane, *Possible Worlds and Other Papers* (1927)

My goal is simple. It is a complete understanding of the universe, why it is as it is and why it exists at all.

> Stephen Hawking, in the *Washington Post*, April 15, 1988

. . . certain conditions under which the observable thing is perceived are tacitly assumed . . . for the possibility that we deal with hallucinations or a dream can never be excluded.

> Ernest H. Hutten, *The Language of Modern Physics* (1956)

"

. . . physics tries to discover the pattern of events which controls the phenomena we observe. But we can never know what this pattern means or how it originates; and even if some superior intelligence were to tell us, we should find the explanation unintelligible.

Sir James Jeans, *Physics and Philosophy* (1958)

We can never finally know. I simply believe that some part of the human Self or Soul is not subject to the laws of space and time.

Carl Jung, in *The Guardian*, July 19, 1975

The highest object at which the natural sciences are constrained to aim, but which they will never reach, is the determination of the forces which are present in nature, and of the state of matter at any given moment—in one word, the reduction of all the phenomena of nature to mechanics.

Gustav Robert Kirchhoff, *Über das Ziel der Naturwissenschaften* (1865)

Given for one instant an intelligence which could comprehend all the forces by which nature is animated and the respective positions of the beings which compose it, if moreover this intelligence were vast enough to submit these data to analysis, it would embrace in the same formula both the movements of the largest bodies in the universe and those of the lightest atom; to it nothing would be uncertain, and the future as the past would be present to its eyes.

Pierre Simon de Laplace, *Oeuvres*, Volume VII,
Théorie Analytique des Probabilités (1812–20)

Truth is a dangerous word to incorporate within the vocabulary of science. It drags with it, in its train, ideas of permanence and immutability that are foreign to the spirit of a study that is essentially an historically changing movement, and that relies so much on practical examination within restricted circumstances. . . . Truth is an absolute notion that science, which is not concerned with any such permanency, had better leave alone.

Hyman Levy, *The Universe of Science* (1933)

Do there exist many worlds, or is there but a single world? This is one of the most noble and exalted questions in the study of nature.

Albertus Magnus, in *A Dictionary of Scientific Quotations*
by Alan L. Mackay (1991)

The world of learning is so broad, and the human soul is so limited in power! We reach forth and strain every nerve, but we seize only a bit of the curtain that hides the infinite from us.

Maria Mitchell, in *Life, Letters, and Journals* (1896)

You geneticists may know something about the hereditary mechanisms that distinguish a red-eyed from a white-eyed fruit fly but you haven't the slightest inkling about the hereditary mechanism that distinguishes fruit flies from elephants.

W. J. Osterhout, 1925, in *Corn, Its Origin, Evolution and Improvement* by P. C. Mangelsdorf (1974)

The universe is full of magical things patiently waiting for our wits to grow sharper.

Eden Phillpotts, in *The World within the World* by John D. Barrow (1988)

Our knowledge can only be finite, while our ignorance must necessarily be infinite.

Sir Karl Popper, *Conjectures and Refutations*

Would there be this eternal seeking if the found existed?

Antonio Porchia, *Voces* (1968)

We are ignorant of the Beyond because this ignorance is the condition *sine qua non* of our own life. Just as ice cannot know fire except by melting, by vanishing.

Jules Renard, Journal, September 1890

The astronomer who studies the motion of the stars is surely like a blind man who, with only a staff [mathematics] to guide him, must make a great, endless, hazardous journey that winds through innumerable desolate places. What will be the result? Proceeding anxiously for a while and groping his way with his staff, he will at some time, leaning upon it, cry out in despair to Heaven, Earth and all the Gods to aid him in his misery.

Georg Joachin Rheticus, in *The Sleepwalkers*
by Arthur Koestler (1968)

Nature does not deceive us; it is we who deceive ourselves.

Jean-Jacques Rousseau, *Emile* (1762)

There is a lurking fear that some things are "not meant" to be known, that some inquiries are too dangerous for human beings to make.

Carl Sagan, *Broca's Brain: Reflections of the Romance of Science* (1979)

The greatest mystery is why there is something instead of nothing, and the greatest something is this thing we call life.

Allan R. Sandage, in *Through a Window*
by Alan Lightman and Roberta Brawer (1990)

"

Some problems are just too complicated for rational logical solutions. They admit of insights, not answers.

> Jerome Bert Wiesner, in "Profiles: A Scientist's Advice II"
> by D. Lang, *The New Yorker*, January 26, 1963

The green pre-human earth is the mystery we were chosen to solve, a guide to the birthplace of our spirit.

> Edward O. Wilson, *The Diversity of Life* (1992)

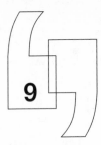

9

Science and Society

That which today calls itself science gives us more and more information, an indigestible glut of information, and less and less understanding.

> Edward Abbey, "Down the River with Henry Thoreau,"
> *Words from the Land* (1981)

If the human race wants to go to hell in a basket, technology can help it get there by jet. It won't change the desire or the direction, but it can greatly speed the passage.

> Charles M. Allen, in a speech at Wake Forest University,
> Winston-Salem, North Carolina, April 25, 1967

The scientists speak with an authority which the ordinary citizen, the non-scientist, cannot challenge, and to which he is compelled to listen. Since they cannot hope for much help from the generals or the ministers, they must act for themselves, in a supreme endeavor to avert the mortal dangers which confront mankind.

Philip Noel Baker, "Science and Disarmament," *Impact,* 1965

. . . science was the great instrument for social change, all the greater because its object is not change but knowledge and its silent appropriation of this dominant function, amid the din of political and religious strife, is the most vital of all the revolutions which have marked the development of modern civilization.

Arthur Balfour, *Decadence* (1908)

Our problems lie not in the genes of the common man but in the ambitions of those with power.

Victor Barnouw, *An Introduction to Anthropology: Volume 2, Ethnology* (1971)

Science, which gave us this dread power, shows that it *can* be made a giant help to humanity, but science does *not* show us how to prevent its baleful use. So we have been appointed to obviate that peril by finding a meeting of the minds and the hearts of our people. Only in the will of mankind lies the answer.

Bernard Baruch, in a plan presented to the U.N. Atomic Energy Commission, June 14, 1946

For the educated, the authority of science rested on the strictness of its method; for the mass, it rested on its powers of explanation.

Jacques Barzun, *Science:
The Glorious Entertainment* (1964)

It is not clear to anyone, least of all the practitioners, how science and technology in their headlong course do or should influence ethics and law, education and government, art and social philosophy, religion and the life of the affections. Yet science is an all-pervasive energy, for it is at once a mode of thought, a source of strong emotion, and a faith as fanatical as any in history.

Jacques Barzun, *Science:
The Glorious Entertainment* (1964)

Fields of learning are surrounded ultimately only by illusory boundaries—like the "rooms" in a hall of mirrors. It is when the illusion is penetrated that progress takes place. . . . Likewise science cannot be regarded as a thing apart, to be studied, admired or ignored. It is a vital part of our culture, our culture is part of it, it permeates our thinking, and its continued separateness from what is fondly called "the humanities" is a preposterous practical joke on all thinking men.

William S. Beck, *Modern Science and
the Nature of Life* (1957)

"

Paradoxically, science . . . has produced institutions, systems of thought and eventually social-political programs that bind people even more than the "superstitions" they replaced.

> Peter L. Berger and Hansfried Kellner,
> *Sociology Reinterpreted* (1981)

Once regarded as the herald of enlightenment in all spheres of knowledge, science is now increasingly seen as a strictly instrumental system of control. Its use as a means of social manipulation and its role in restricting human freedom now parallel in every detail its use as a means of natural manipulation.

> Murray Bookchin, *The Ecology of Freedom* (1982)

Every thoughtful man who hopes for the creation of a contemporary culture knows that this hinges on one central problem: to find a coherent relation between science and the humanities.

> Jacob Bronowski and Bruce Mazlish,
> *The Western Intellectual Tradition* (1960)

There is virtually nothing that has come from molecular biology that can be of any value to human living in the conventional sense of what is good, and quite tremendous possibilities of evil, again in the conventional sense.

> Sir Frank Macfarlane Burnet, *Changing Patterns* (1968)

"

The sum of human happiness would not necessarily be reduced if for ten years every physical and chemical laboratory were closed and the patient and resourceful energy displayed in them transferred to the lost art of getting on together and finding the formula for making both ends meet in the scale of human life.

> Bishop Edward Arthur Burroughs, in a speech to the
> British Association for the Advancement of Science,
> Leeds, September 4, 1927

[The Scientific Revolution] outshines everything since the rise of Christianity and reduces the Renaissance and the Reformation to the rank of mere episodes, mere internal displacements, within the system of medieval Christendom. . . . It looms so large as the real origin of the modern world and of the modern mentality that our customary periodization of European history has become an anachronism and an encumbrance.

> Herbert Butterfield, *The Origins of Modern Science* (1949)

You don't ask the frogs when you drain the marsh.

> Rémy Louis Carle, a director of Électricité de France,
> when asked if the public was consulted on the siting of a
> nuclear power station, in *The Guardian,* 1986

There is no national science just as there is no national multiplication table; what is national is no longer science.

> Anton Chekhov, in *Mysli o Nauke* by V. P. Ponomarev (1973)

"

By God's mercy British and American science outpaced all German efforts. . . . This revelation of the secrets of nature, long mercifully withheld from man, should arouse the most solemn reflections in the mind and conscience of every human being capable of comprehension. We must indeed pray that these awful agencies will be made to conduce to peace among the nations, and that instead of wreaking measureless havoc upon the entire globe, they may become a perennial fountain of world prosperity.

Winston Churchill, *Statements Relating to the Atomic Bomb* (1945)

Science, which now offers us a golden age with one hand, offers at the same time with the other the doom of all that we have built up inch by inch since the Stone Age and the dawn of any human annals. My faith is in the high progressive destiny of man. I do not believe we are to be flung back into abysmal darkness by those fiercesome discoveries which human genius has made. Let us make sure that they are servants, but not our masters.

Winston Churchill, in *The Wit & Wisdom of Winston Churchill* by James C. Humes (1994)

The spiritual and intellectual decline which has overtaken us in the last thirty years . . . [may be due] to the diversion of all the best brains to technology.

Lord Kenneth Clarke, in *London Review of Books,* 1984

Science is triumphant with far-ranging success, but its triumph is somehow clouded by growing difficulties in providing for the simple necessities of human life on earth.

Barry Commoner, *Science and Survival* (1966)

In recent times, modern science has developed to give mankind, for the first time in the history of the human race, a way of securing a more abundant life which does not simply consist in taking away from someone else.

Karl Taylor Compton, in a speech to the American Philosophical Society, 1938

Take the so called standard of living. What do most people mean by "living"? They don't mean living. They mean the latest and closest plural approximation to singular prenatal passivity which science, in its finite but unbounded wisdom, has succeeded in selling their wives.

e. e. cummings, Introduction, *Poems* (1954)

Accountants and second-rate business school jargon are in the ascendant. Costs, which rise rapidly, and are easily ascertained and comprehensible, now weigh more heavily in the scales than the unquantifiable and unpredictable values and future material progress. Perhaps science will only regain its lost primacy as peoples and government begin to recognize that sound scientific work is the only secure basis

"

for the construction of policies to ensure the survival of Mankind without irreversible damage to Planet Earth.

Sir Frederick Sydney Dainton,
in *New Scientist,* March 3, 1990

The origin of what we call civilization is not due to religion but to skepticism. . . . The modern world is the child of doubt and inquiry, as the ancient world was the child of fear and faith.

Clarence Darrow, in *Summer for the Gods*
by Edward J. Larson (1997)

Most of the dangerous aspects of technological civilization arise, not from its complexities, but from the fact that modern man has become more interested in the machines and industrial goods themselves than in their use to human ends.

René Dubos, *A God Within* (1972)

The contents of physics is the concern of physicists, its effect the concern of all men.

Friedrich Dürrenmatt, "21 Points," *The Physicists* (1962)

There are three reasons why, quite apart from scientific considerations, mankind needs to travel in space. The first reason is garbage disposal; we need to transfer industrial processes into space so that the earth may remain a green and pleasant place for our grandchildren to live in. The second reason is to escape material impoverishment; the resources of this planet are finite, and we shall not forgo forever the abundance of solar energy and minerals and living space that are spread out all around us. The third reason is our spiritual need for an open frontier. The ultimate purpose of space travel is to bring to humanity, not only scientific discoveries and an occasional spectacular show on television, but a real expansion of our spirit.

Freeman J. Dyson, *Disturbing the Universe* (1979)

The technologies which have had the most profound effects on human life are usually simple. A good example of a simple technology with profound historical consequences is hay. Nobody knows who invented hay, the idea of cutting grass in the autumn and storing it in large enough quantities to keep horses and cows alive through the winter. All we know is that the technology of hay was unknown to the Roman Empire but was known to every village of medieval Europe. Like many other crucially important technologies, hay emerged anonymously during the so-called Dark Ages. According to the Hay Theory of History, the invention of hay was the decisive event which moved the center of gravity of urban civilization from the Mediterranean basin to Northern and Western Europe. The Roman Empire did not need hay because in a Mediterranean climate the grass grows well enough in winter for animals to graze. North of the Alps, great cities dependent on horses and oxen for

motive power could not exist without hay. So it was hay
that allowed populations to grow and civilizations to
flourish among the forests of Northern Europe. Hay moved
the greatness of Rome to Paris and London, and later to
Berlin and Moscow and New York.

Freeman J. Dyson, *Infinite in All Directions* (1988)

Concern for man and his fate must always form the chief
interest of all technical endeavors . . . in order that the
creations of our mind shall be a blessing and not a curse to
mankind. Never forget this in the midst of your diagrams
and equations.

Albert Einstein, in a speech at the California Institute of
Technology, Pasadena, February 1931

Why does this magnificent applied science which saves
work and makes life easier bring us so little happiness? The
simple answer runs: Because we have not yet learned to
make sensible use of it.

Albert Einstein, in a speech at the California Institute of
Technology, Pasadena, February 1931

The release of atomic energy has not created a new
problem. It has merely made more urgent the necessity
of solving an existing one.

Albert Einstein, "Atomic War or Peace,"
Atlantic Monthly, November 1945

Responsibility lies with those who make use of these new tools and not with those who contribute to the progress of knowledge: therefore, with the politicians, not with the scientists.

> Albert Einstein, discussing atomic weapons,
> in an interview, February 1949

That is simple, my friend: because politics is more difficult than physics.

> Albert Einstein, when asked why people could
> discover atoms but not the means to control them,
> in *The New York Times,* April 22, 1955

In my opinion it is not right to bring politics into scientific matters, nor should individuals be held responsible for the government of the country to which they happen to belong.

> Albert Einstein, in *Einstein: A Centenary Volume*
> edited by A. P. French (1979)

One must divide one's time between politics and equations. But our equations are much more important to me.

> Albert Einstein, in *Einstein: A Centenary Volume*
> edited by A. P. French (1979)

It is frequently the tragedy of the great artist, as it is of the great scientist, that he frightens the ordinary man.

> Loren Eiseley, *The Night Country* (1971)

One day, Sir, you may tax it.

> Michael Faraday, answering the Chancellor of the Exchequer's question about the practical value of electricity, in *Discovery of the Spirit and Service of Science* by R. A. Gregory (1916)

One of the great problems of the world today is undoubtedly this problem of not being able to talk to scientists, because we don't understand science; they can't talk to us because they don't understand anything else, poor dears.

> Michael Flanders, *At the Drop of a Hat* (1964)

Science has radically changed the conditions of human life on earth. It has expanded our knowledge and our power but not our capacity to use them with wisdom.

> Senator J. William Fulbright,
> *Old Myths and New Realities* (1964)

. . . a science is said to be useful if its development tends to accentuate the existing inequalities in the distribution of wealth, or more directly promotes the destruction of human life.

> Harold Hardy Godfrey, *A Mathematician's Apology* (1941)

Once upon a time we were just plain people. But that was before we began having relationships with mechanical systems. Get involved with a machine and sooner or later you are reduced to a factor.

Ellen Goodman, "The Human Factor,"
The Washington Post, January 1987

Man and science are two concave mirrors continually reflecting each other.

Aleksandr Ivanovich Herzen, *Science and Humanity* (1968)

There is not much that even the most socially responsible scientists can do as individuals, or even as a group, about the social consequences of their activities.

Eric John Ernest Hobsbawm,
in *New York Review of Books*, November 19, 1970

To this day, we see all around us the Promethean drive to *omnipotence through technology* and to *omniscience through science*. The effecting of all things possible and the knowledge of all causes are the respective primary imperatives of technology and of science. But the motivating imperative of society continues to be the very different one of its physical and spiritual survival. It is not far less obvious than it was in Francis Bacon's world how to bring the three imperatives into harmony, and how to bring all three together to bear on problems where they superpose.

Gerald Holton, "Science, Technology, and
the Fourth Discontinuity," *The Advancement
of Science, and Its Burdens* (1986)

"

All the sciences have a relation, greater or less, to human nature; and . . . however wide any of them may seem to run from it, they still return back by one passage or another. Even *Mathematics, Natural Philosophy, and Natural Religion,* are in some measure dependent on the science of Man; since they lie under the cognizance of men, and are judged by their powers and faculties.

David Hume, *A Treatise on Human Nature* (1739–40)

The cosmogonist has finished his task when he has described to the best of his ability the inevitable sequence of changes which constitute the history of the material universe. But the picture which he draws opens questions of the widest interest not only to science, but also to humanity. What is the significance of the vast processes it portrays? What is the meaning, if any there be which is intelligible to us, of the vast accumulations of matter which appear, on our present interpretations of space and time, to have been created only in order that they may destroy themselves.

Sir James Jeans, *Astronomy and Cosmogony* (1961)

We have genuflected before the god of science only to find that it has given us the atomic bomb, producing fears and anxieties that science can never mitigate.

Martin Luther King, Jr., *Strength to Love* (1963)

Science frees us in many ways . . . from the bodily terror
which the savage feels. But she replaces that, in the minds
of many, by a moral terror which is far more overwhelming.

Charles Kingsley, in a sermon, November 26, 1866

Nominally a great age of scientific inquiry, ours has actually
become an age of superstition about the infallibility of
science; of almost mystical faith in its nonmystical methods;
above all . . . of external verities; of traffic-cop morality
and rabbit-test truth.

Louis Kronenberger, *Company Manners* (1954)

Science has always promised two things not necessarily
related—an increase first in our powers, second in our
happiness and wisdom, and we have come to realize that it
is the first and less important of the two promises which it
has kept most abundantly.

Joseph Wood Krutch, "The Disillusion with the Laboratory,"
The Modern Temper (1929)

It might be going too far to say that the modern scientific
movement was tainted from its birth; but I think it would
be true to say that it was born in an unhealthy neighborhood
and at an inauspicious hour. Its triumphs may have been
too rapid and purchased at too high a price: reconsideration,
and something like repentance, may be required.

C. S. Lewis, *The Abolition of Man* (1978)

"

The form of society has a very great effect on the rate of inventions and a form of society which in its young days encourages technical progress can, as a result of the very inventions it engenders, eventually come to retard further progress until a new social structure replaces it. The converse is also true. Technical progress affects the structure of society.

Sam Lilley, *Men, Machines and History* (1948)

I have seen the science I worshipped, and the aircraft I loved, destroying the civilization I expected them to serve.

Charles A. Lindbergh, in *Time*, May 26, 1967

Marxist philosophy holds that the most important problem does not lie in understanding the laws of the objective world and thus being able to explain it, but in applying the knowledge of these laws actively to change the world.

Mao Tse-Tung, "On Practice,"
Quotations from Chairman Mao Tse-Tung (1967)

Natural science is one of man's weapons in his fight for freedom. For the purpose of attaining freedom in society, man must use social science to understand and change society and carry out social revolution. For the purpose of attaining freedom in the world of nature, man must use natural science to understand, conquer and change nature and thus attain freedom from nature.

> Mao Tse-Tung, in a speech to the Natural Science Research
> Society for the Border Regions, *Quotations from
> Chairman Mao Tse-Tung* (1967)

History itself is an actual part of natural history, of nature's development into man. Natural science will in time include the science of man as the science of man will include natural science: there will be one science.

> Karl Marx, *Writings of the Young Marx on Philosophy and
> Society* edited by L. D. Easton and K. H. Guddat (1967)

Science and technology, and the various forms of art, all unite humanity in a single and interconnected system. As science progresses, the worldwide cooperation of scientists and technologists becomes more and more of a special and distinct intellectual community of friendship, in which, in place of antagonism, there is growing up a mutually advantageous sharing of work, a coordination of efforts, a common language for the exchange of information, and a solidarity, which are in many cases independent of the social and political differences of individual states.

> Zhores Aleksandrovich Medvedev, *The Medvedev Papers* (1970)

Armed with all the powers, enjoying all the riches they owe to science, our societies are still trying to live by and to

"

teach systems of values already blasted at the root by science itself.

Jacques Monod, *Chance and Necessity* (1970)

Science—we have loved her well, and followed her diligently, what will she do? I fear she is too much in the pay of the counting-houses, and the drill-serjent, that she is too busy, and will for the present do nothing. Yet there are matters which I should have thought easy for her; say, for example, teaching Manchester how to consume its town smoke, or Leeds how to get rid of its superfluous black dye without turning it into the river, which would be as much worth her attention as the production of the heaviest black silks, or the biggest of useless guns.

William Morris, *The Lesser Arts* (1878)

In ancient days heroes slew the dragons that terrified men; science is destroying the more dreadful superstitions that have darkened human lives.

Forest Ray Moulton and Justus J. Schifferes,
The Autobiography of Science (Second Edition) (1960)

What it is important to realize is that automation . . . is an attempt to exercise control, not only of the mechanical process itself, but of the human being who once directed it:

turning him from an active to a passive agent, and finally eliminating him all together.

> Lewis Mumford, "The Myth of the Machine,"
> *The Pentagon of Power* (1970)

Democracy might therefore almost in a sense be termed that practice of which science is the theory.

> John Needham, *The Grand Titration* (1969)

It is science alone that can solve the problems of hunger and poverty, insanitation and illiteracy, of superstition and deadening custom and tradition, of vast resources running to waste, of a rich country inhabited by starving people. . . . Who indeed could afford to ignore science today? At every turn we have to seek its aid. . . . The future belongs to science and to those who make friends with science.

> Jawaharlal Nehru, in *Proceedings of the*
> *National Institute of Sciences of India,* 1961

Modern technology has lost its magic. No longer do people stand in awe, thrilled by the onward rush of science, the promise of a new day. Instead, the new is suspect. It arouses our hostility as much as it used to excite our fancy. With each breakthrough there are recurrent fears and suspicion. How will the advance further pollute our lives; modern technology is not merely what it first appears to be. Behind the white coats, the disarming jargon, the elaborate instrumentation, and at the core of what has often seemed an automatic process, one finds what Dorothy found in Oz: modern technology is human after all.

David Noble, in *Science and Liberation*
edited by Rita Arditti, Pat Brennan, and Steve Cavrak (1980)

The scientist is not responsible for the laws of nature, but it is a scientist's job to find out how these laws operate. It is the scientist's job to find ways in which these laws can serve the human will. However, it is not the scientist's job to determine whether a hydrogen bomb should be used. . . .

J. Robert Oppenheimer, in *A Passion for Science*
edited by L. Wolpert and A. Richards (1988)

Science is intimately integrated with the whole social structure and cultural tradition. They mutually support each other—only in certain types of society can science flourish, and conversely without a continuous and healthy development and application of science such a society cannot function properly.

Talcott Parsons, *The Social System* (1951)

Science knows no country, because knowledge belongs to humanity, and is the torch which illuminates the world. Science is the highest personification of the nation because that nation will remain the first which carries the further the works of thought and intelligence.

Louis Pasteur, in *Pasteur and Modern Science*
by René Dubos (1960)

Science is the search for truth—it is not a game in which one tries to beat his opponents, to do harm to others. We need to have the spirit of science in international affairs, to make the conduct of international affairs the effort to find the right solution, the just solution of international problems, not the effort by each nation to get the better of other nations. . . .

Linus Pauling, *No More War!* (1958)

The major producer of the social chaos, the indeterminacy of thought and values that rational knowledge is supposed to eliminate, is none other than science itself.

Robert M. Pirsig, *Zen and the Art of
Motorcycle Maintenance* (1974)

The way to solve the conflict between human values and technological needs is not to run away from technology, that's impossible. The way to resolve the conflict is to break down the barriers of dualistic thought that prevent a real understanding of what technology is—not an exploitation

"

of nature, but a fusion of nature and the human spirit into a new kind of creation that transcends both.

> Robert M. Pirsig, *Zen and the Art of*
> *Motorcycle Maintenance* (1974)

Science . . . cannot exist on the basis of a treaty of strict non-aggression with the rest of society; from either side, there is no defensible frontier.

> Don Krasher Price, *Government and Science* (1954)

The general public has long been divided into two parts; those who think science can do anything, and those who are afraid it will.

> Dixie Lee Ray, in *New Scientist,* July 5, 1973

While government and laws provide for the safety and well-being of assembled men, the sciences, letters and arts, less despotic and perhaps more powerful, spread garlands of flowers over the iron chains with which men are burdened, stifle in them the sense of that original liberty for which they seemed to have been born, make them love their slavery, and turn them into what is called civilized peoples.

> Jean-Jacques Rousseau,
> *Discourse on the Sciences and Arts* (1750)

He who makes two blades of grass grow where one grew before is the benefactor of mankind, but he who obscurely worked to find the laws of such growth is the intellectual superior as well as the greater benefactor of mankind.

> Henry Augustus Rowland, in *The Politics of Pure Science*
> by D. S. Greenberg (1967) [See Jonathan Swift]

I am compelled to fear that science will be used to promote the power of dominant groups rather than to make men happy.

> Bertrand Russell, *Icarus, or the Future of Science* (1925)

It is essential for men of science to take an interest in the administration of their own affairs or else the professional civil servant will step in—and then the Lord help you.

> Lord Ernest Rutherford,
> in *Bulletin of the Institute of Physics,* 1950

The intellectual life of the whole of western society is increasingly being split into two polar groups. . . . Literary intellectuals at one pole—at the other scientists. . . . Between the two a gulf of mutual incomprehension.

> C. P. Snow, *The Two Cultures and*
> *the Scientific Revolution* (1959)

The hype, skepticism and bewilderment associated with the Internet—concerns about new forms of crime, adjustments in social mores, and redefinition of business practices— mirror the hopes, fears, and misunderstandings inspired by

"

the telegraph. Indeed, they are only to be expected. They are the direct consequences of human nature, rather than technology.

Given a new invention, there will always be some people who see only its potential to do good, while others see new opportunities to commit crime or make money. We can expect the same reactions to whatever new inventions appear in the twenty-first century.

Such reactions are amplified by what might be termed chronocentricity—the egotism that one's own generation is poised on the very cusp of history. Today, we are repeatedly told that we are in the midst of a communications revolution. But the electric telegraph was, in many ways, far more disconcerting for the inhabitants of the time than today's advances are for us. If any generation has the right to claim that it bore the full bewildering, world-shrinking brunt of such a revolution, it is not us—it is our nineteenth-century forebears.

Tom Standage, *The Victorian Internet* (1998)

With the unlocking of the atom, mankind crossed one of the great watersheds of history. We have entered uncharted lands. The maps of strategy and diplomacy by which we guided ourselves until yesterday no longer reveal the way. Fusion and fission revolutionized the whole foundation of human affairs.

Adlai E. Stevenson, in a speech to the General Federation of Women's Clubs, Philadelphia, Pennsylvania, May 24, 1955

Science is our favorite modern superstition.

> Thomas Stirton, in conversation, 1970

. . . whoever could make two ears of corn, or two blades of grass, to grow upon a spot of ground where only one grew before, would deserve better of mankind, and do more essential service to this country, than the whole race of politicians put together.

> Jonathan Swift, *Gulliver's Travels* (1726)
> [See Henry Augustus Rowland]

Basic research may seem very expensive. I am a well-paid scientist. My hourly wage is equal to that of a plumber, but sometimes my research remains barren of results for weeks, months or years and my conscience begins to bother me for wasting the taxpayer's money. But in reviewing my life's work, I have to think that the expense was not wasted. Basic research, to which we owe everything, is relatively very cheap when compared with other outlays of modern society. The other day I made a rough calculation which led me to the conclusion that if one were to add up all the money ever spent by man on basic research, one would find it to be just about equal to the money spent by the Pentagon this past year.

> Albert Szent-Györgyi, *The Crazy Ape* (1971)

If the arrangement of society is bad (as ours is), and a small number of people have power over the majority and oppress it, every victory over Nature will inevitably serve only to increase that power and that oppression.

Leo Tolstoy, in *Science, Liberty and Peace*
by Aldous Huxley (1947)

But science and technology are only one of the avenues toward reality; others are equally needed to comprehend the full significance of our existence. Indeed, these other avenues are necessary for the prevention of thoughtless and inhuman abuses of the results of science.

Victor Frederick Weisskopf,
The Privilege of Being a Physicist (1989)

The value of fundamental research does not lie only in the ideas it produces. There is more to it. It affects the whole intellectual life of a nation by determining its way of thinking and the standards by which actions and intellectual production are judged. If science is highly regarded and if the importance of being concerned with the most up-to-date problems of fundamental research is recognized, then a spiritual climate is created which influences the other activities. An atmosphere of creativity is established which penetrates every cultural frontier. Applied sciences and technology are forced to adjust themselves to the highest intellectual standards which are developed in the basic sciences. This influence works in many ways: some fundamental students go into industry;

the techniques which are applied to meet the stringent requirements of fundamental research serve to create new technological methods. The style, the scale, and the level of scientific and technical work are determined in pure research; that is what attracts productive people and what brings scientists to those countries where science is at the highest level. Fundamental research sets the standards of modern scientific thought; it creates the intellectual climate in which our modern civilization flourishes. It pumps the lifeblood of idea and inventiveness not only into the technological laboratories and factories, but into every cultural activity of our time. The case for generous support for pure and fundamental science is as simple as that.

Victor Frederick Weisskopf, "Why Pure Science?" in *Bulletin of the Atomic Scientists*, 1965

Why does not science, instead of troubling itself about sunspots, which nobody ever saw, or, if they did, ought not to speak about; why does not science busy itself with drainage and sanitary engineering? Why does it not clean the streets and free the rivers from pollution?

Oscar Wilde, in *Oscar Wilde: Interviews and Recollections* by E. H. Mikhail (1979)

"

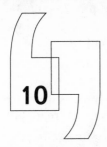

10

The Science of Mathematics, the Mathematics of Science

. . . Like the statistician who drowned in a lake of average depth six inches.

> Anonymous, in *A Dictionary of Scientific Quotations*
> by Alan L. Mackay (1991)

Mathematics is the door and the key to the sciences.

> Roger Bacon, *Opus Majus* (1267)

If a lunatic scribbles a jumble of mathematical symbols it does not follow that the writing means anything merely because to the inexpert eye it is indistinguishable from higher mathematics.

> Eric Temple Bell, in *The World of Mathematics*
> edited by J. R. Newman (1956)

Calculation touches, at most, certain phenomena of organic destruction. Organic creation, on the contrary, the evolutionary phenomena which properly constitute life, we cannot in any way subject to a mathematical treatment.

Henri Bergson, *Creative Evolution* (1911)

A witty statesman said, you might prove anything by figures.

Thomas Carlyle, *Chartism* (1840)

Some of my cousins who had the great advantage of University education used to tease me with arguments to prove that nothing has any existence except what we think of it. . . . These amusing mental acrobatics are all right to play with. They are perfectly harmless and perfectly useless. . . . I always rested on the following argument. . . . We look up to the sky and see the sun. Our eyes are dazzled and our senses record the fact. So here is this great sun standing apparently on no better foundation than our physical senses. But happily there is a method, apart altogether from our physical senses, of testing the reality of the sun. It is by mathematics. By means of prolonged processes of mathematics, entirely separate from the senses, astronomers are able to calculate when an eclipse will occur. They predict by pure reason that a black spot will pass across the sun on a certain day. You go and look, and your sense of sight immediately tells you that their calculations are vindicated. So here you have the evidence of the senses reinforced by the entirely separate evidence of a vast independent process of mathematical reasoning. We have taken what is called in military map-making "a cross bearing." When my metaphysical friends tell me that the data on which the astronomers made their calculations, were necessarily obtained originally through the evidence of the senses, I say, "no." They might, in theory at any rate, be obtained by automatic calculating-machines set in motion by the light falling upon them without admixture of the human senses at any stage. When it is persisted that we should have to be told about the calculations and use our ears for that purpose, I reply that the mathematical process has a reality and virtue in itself, and that once discovered it constitutes a new and independent factor. I am also at this point accustomed to reaffirm with emphasis my conviction that the sun is real, and also that it is hot— in fact hot as Hell, and that if the metaphysicians doubt it they should go there and see.

Winston Churchill, My Early Life (1930)

No human investigation can be called real science if it cannot be demonstrated mathematically.

Leonardo da Vinci, Treatise on Painting (1651)

One reason why mathematics enjoys special esteem, above all other sciences, is that its laws are absolutely certain and indisputable, while those of all other sciences are to some extent debatable and in constant danger of being overthrown by newly discovered facts.

> Albert Einstein, *Sidelights on Relativity* (1922)

I don't believe in mathematics.

> Albert Einstein, in *Albert Einstein* by Paul Seelig (1956)

As far as the laws of mathematics refer to reality, they are not certain, and as far as they are certain, they do not refer to reality.

> Albert Einstein, in *The Tao of Physics* by Fritjof Capra (1975)

The mathematician has reached the highest rung on the ladder of human thought.

> Havelock Ellis, *The Dance of Life* (1923)

It does not follow that because something *can* be counted it therefore *should* be counted.

> Harold Enarson, in a speech to the Society for College and University Planning, September 1975

Physics is to mathematics what sex is to masturbation.

> Richard P. Feynman, in *Fear of Physics* by Lawrence M. Krauss (1993)

Philosophy is written in that great book which ever lies before our gaze—I mean the universe—but we cannot understand it if we do not first learn the language and grasp the symbols in which it is written. The book is written in the mathematical language, and the symbols are triangles, circles and other geometrical figures, without the help of which it is impossible to conceive a single word of it, and without which one wanders in vain through a dark labyrinth.

Galileo Galilei, *Opere Il Sattiatore* (1656)

Beauty is the first test; there is no permanent place in the world for ugly mathematics.

Godfrey Harold Hardy, *A Mathematician's Apology* (1941)

The Universe can be pictured, although still very imperfectly and inadequately, as consisting of pure thought, the thought of what for want of a wider word, we must describe as a mathematical thinker.

Sir James Jeans, *The Mysterious Universe* (1948)

The chief aim of all investigations of the external world should be to discover the rational order and harmony which has been imposed on it by God and which He revealed to us in the language of mathematics.

Johannes Kepler, in *Mathematical Thought from Ancient to Modern Times* by Morris Kline (1972)

"

All the effects of nature are only the mathematical consequences of a small number of immutable laws.

> Pierre Simon de Laplace, in *Men of Mathematics* by E. T. Bell (1965)

Stand firm in your refusal to remain conscious during algebra. In real life, I assure you, there is no such thing as algebra.

> Fran Lebowitz, "Tips for Teens," *Social Studies* (1981)

Man knows only these poor mathematical theories about the heavens, and only God knows the real motions of the heavens and their causes.

> Moses Maimonides, in *Modern Science and Its Philosophy* by Philip Frank (1950)

Today, it is not only that our kings do not know mathematics, but our philosophers do not know mathematics and— to go a step further—our mathematicians do not know mathematics.

> J. Robert Oppenheimer, "The Tree of Knowledge," *Harper's*, 1958

The more progress physical sciences make, the more they tend to enter the domain of mathematics, which is a kind of center to which they all converge. We may even judge of the degree of perfection to which a science has arrived by the facility with which it may be submitted to calculation.

Adolphe Quetelet, in *Eulogy of Quetelet* by E. Mailly (1874)

Mathematics may be defined as the subject in which we never know what we are talking about, nor whether what we are saying is true.

Bertrand Russell, "Mathematics and the Metaphysicians," *Mysticism and Logic* (1917)

Mathematics, rightly viewed, possesses not only truth, but supreme beauty—a beauty cold and austere, like that of a sculpture, without appeal to any part of our weaker nature, without the gorgeous trappings of painting or music, yet sublimely pure, and capable of a stern perfection such as only the greatest art can show.

Bertrand Russell, "The Study of Mathematics," *Mysticism and Logic* (1917)

Ordinary language is totally unsuited for expressing what physics really asserts, since the words of everyday life are not sufficiently abstract. Only mathematics and mathematical logic can say as little as the physicist means to say.

Bertrand Russell, *The Scientific Outlook* (1931)

"

Physics is mathematical not because we know so much about the physical world, but because we know so little: it is only its mathematical properties that we can discover.

> Bertrand Russell, in The *World within the World*
> by John D. Barrow (1988)

If your experiment needs statistics, you ought to have done a better experiment.

> Henry Norris Russell, in *The Mathematical Approach
> to Biology and Medicine* by N. T. J. Bailey (1967)

I often say that when you can measure what you are speaking about and express it in numbers you know something about it; but when you cannot measure it, when you cannot express it in numbers, your knowledge is of a meager and unsatisfactory kind: it may be the beginning of knowledge, but you have scarcely, in your thoughts, advanced to the stage of science, whatever the matter may be.

> William Thomson (Lord Kelvin), in a lecture at
> the Institution of Civil Engineers, May 3, 1883

Do not imagine that mathematics is hard and crabbed, and repulsive to common sense. It is merely the etherialization of common sense.

> William Thomson (Lord Kelvin), in *Life of Lord Kelvin*
> by S. P. Thompson (1910)

In mathematics you don't understand things. You just get used to them.

> Johann von Neumann, in *The Dancing Wu Li Masters*
> by Gary Zukav (1979)

It is a safe rule to apply that, when a mathematical or philosophical author writes with a misty profundity, he is talking nonsense.

> Alfred North Whitehead,
> *An Introduction to Mathematics* (1911)

"

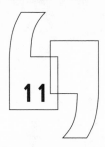

11

Ambition, Success, and Failure

Part of the strength of science is that it has tended to attract
individuals who love knowledge and the creation of it.
Just as important to the integrity of science have been the
unwritten rules of the game. These provide recognition and
approbation for work which is imaginative and accurate,
and apathy or criticism for the trivial or inaccurate. . . .
Thus, it is the communication process which is at the core
of the vitality and integrity of science. . . . The system of
rewards and punishments tends to make honest, vigorous,
conscientious, hardworking scholars out of people who
have human tendencies of slothfulness and no more rectitude
than the law requires.

Phillip Hauge Abelson, "The Roots of Scientific Integrity,"
in *Science*, 1963

Sure Prometheus discovered fire, but what has he done since?
Anonymous, in *A Dictionary of Scientific Quotations*
by Alan L. Mackay (1991)

Truth comes out of error more readily than out of confusion.

Sir Francis Bacon, *Novum Organum* (1620)

Anyone of common mental and physical health can practice scientific research. . . . Anyone can try by patient experiment what happens if this or that substance be mixed in this or that proportion with some other under this or that condition. Anyone can vary the experiment in any number of ways. He that hits in this fashion on something novel and of use will have fame. . . . The fame will be the product of luck and industry. It will not be the product of special talent.

Hilaire Belloc, *Essays of a Catholic Layman in England* (1931)

What counts . . . in science is not so much the first as the last.

Erwin Chargaff, *Science,* 1971

It is not easy to convey, unless one has experienced it, the dramatic feeling of sudden enlightenment that floods the mind when the right idea finally clinches into place.

Francis Harry Compton Crick, *What Mad Pursuit* (1988)

When the war finally came to an end, I was at a loss as to what to do. . . . I took stock of my qualifications. A not-very-good degree, redeemed somewhat by my achievements at the Admiralty. A knowledge of certain restricted parts of magnetism and hydrodynamics, neither of them subjects for which I felt the least bit of enthusiasm. No published papers at all . . . [O]nly gradually did I realize that this lack of qualification could be an advantage. By the time most scientists have reached age thirty they are trapped by their own expertise. They have invested so much effort in one particular field that it is often extremely difficult, at that time in their careers, to make a radical change. I, on the other hand, knew nothing, except for a basic training in somewhat old-fashioned physics and mathematics and an ability to turn my hand to new things. . . . Since I essentially knew nothing, I had an almost completely free choice. . . .

Francis Harry Compton Crick, *What Mad Pursuit* (1988)

A few of the results of my activities as a scientist have become embedded in the very texture of the science I tried to serve—this is the immortality that every scientist hopes for. I have enjoyed the privilege, as a university teacher, of being in a position to influence the thought of many hundreds of young people and in them and in their lives I shall continue to live vicariously for a while. All the things I care for will continue for they will be served by those who come after me. I find great pleasure in the thought that those who stand on my shoulders will see much farther than I did in my time. What more could any man want?

Francis Albert Eley Crew, "The Meaning of Death," in *The Humanist Outlook* edited by A. J. Ayer (1968)
[See Gerald Holton and Sir Isaac Newton]

"

In science the credit goes to the man who convinces the world, not to the man to whom the idea first occurs.

Sir Francis Darwin, in Eugenics Review, April 1914

The law that entropy always increases—the Second Law of Thermodynamics—holds, I think, the supreme position among the laws of Nature. If someone points out to you that your pet theory of the universe is in disagreement with Maxwell's equation—then so much the worse for Maxwell's equation. If it is found to be contradicted by observation—well, these experimentalists do bungle things sometimes. But if your theory is found to be against the Second Law of Thermodynamics I can give you no hope; there is nothing for it but to collapse in deepest humiliation.

Sir Arthur Stanley Eddington,
The Nature of the Physical World (1928)

Neither on my deathbed nor before will I ask myself such a question. Nature is not an engineer or a contractor, and I myself am a part of Nature.

Albert Einstein, responding to a question concerning
what would determine the success or failure of his life,
November 12, 1930, in Albert Einstein, the Human Side
by Helen Dukas and Banesh Hoffman (1979)

Try not to become a man of success, but rather try to become a man of value.

Albert Einstein, in Life, May 2, 1955

"

Were I wrong, one professor would have been quite
enough.

> Albert Einstein, in response to a book in which 100 Nazi
> professors charged him with scientific error,
> in *The Washington Post,* December 12, 1978

No amount of experimentation can ever prove me right; a
single experiment can prove me wrong.

> Albert Einstein (attributed)

It was the failures who had always won, but by the time
they won they had come to be called successes. This is the
final paradox, which men call evolution.

> Loren Eiseley, *The Star Thrower* (1978)

An expert is someone who knows some of the worst
mistakes that can be made in his subject, and how to
avoid them.

> Werner Heisenberg, *Physics and Philosophy* (1971)

As a result of the phenomenally rapid change and growth
of physics, the men and women who did their great work
one or two generations ago may be our distant predecessors
in terms of the state of the field, but they are our close
neighbors in terms of time and tastes. This may be an
unprecedented state of affairs among professionals; one can
perhaps be forgiven if one characterizes it epigrammatically
with a disastrously mixed metaphor; in the sciences, we are

"

now uniquely privileged to sit side-by-side with the giants on whose shoulders we stand.

> Gerald Holton, "On the Recent Past of Physics,"
> *American Journal of Physics,* 1961
> [See Sir Isaac Newton and Francis Albert Eley Crew]

The great tragedy of Science—the slaying of a beautiful hypothesis by an ugly fact.

> T. H. Huxley, "Biogenesis and Abiogenesis,"
> *Critiques and Addresses* (1873)

First . . . a new theory is attacked as absurd; then it is admitted to be true, but obvious and insignificant; finally it is seen to be so important that its adversaries claim that they themselves discovered it.

> William James, *Pragmatism* (1907)

The requirement for great success is great ambition . . . for triumph over other men, not merely over nature.

> Richard C. Lewontin, in *Patenting the Sun:
> Polio and the Salk Vaccine* by Jane S. Smith (1990)

I know of nothing so pleasant to minds as the discovery of anything which is at once new and valuable; for nothing which so lightens and sweetens toil, as the hopeful pursuit of such discovery.

> Abraham Lincoln, in *The Wit & Wisdom of Abraham Lincoln*
> by James C. Humes (1996)

Mankind always takes up only such problems as it can
solve; since, looking at the matter more closely, we will
always find that the problem itself arises only when the
material conditions necessary for its solution already exist
or are at least in the process of formation.

> Karl Marx, in *Selected Works of Karl Marx*
> *and Frederick Engels,* Volume One (1962)

I must admit that I personally measure success in terms of
the contributions an individual makes to her or his fellow
human beings.

> Margaret Mead, in *Redbook,* November 1978

No scientist is admired for failing in the attempt to solve
problems that lie beyond his competence. The most he can
hope for is the kindly contempt earned by the Utopian
politician. If politics is the art of the possible, research is
surely the art of the soluble. Good scientists study the most
important problems they think they can solve. It is, after
all, their professional business to solve problems, not
merely to grapple with them.

> Sir Peter Medawar, *The Art of the Soluble* (1967)

There is no spiritual copyright in scientific discoveries,
unless they should happen to be quite mistaken. Only in
making a blunder does a scientist do something which,
conceivably, no one might ever do again.

> Sir Peter Medawar, *Pluto's Republic* (1976)

"

Scientific discovery is a private event, and the delight that accompanies it, or the despair of finding it illusory does not travel.

Sir Peter Medawar, *Hypothesis and Imagination*

The complex pattern of the misallocation of credit for scientific work must quite evidently be described as "the Matthew effect," for, as will be remembered, the Gospel According to St. Matthew puts it this way:

For unto every one that hath shall be given, and he shall have abundance: but from him that hath not shall be taken away even that which he hath.

Put in less stately language, the Matthew effect consists of the accruing of greater increments of recognition for particular scientific contributions to scientists of considerable repute and the withholding of such recognition from scientists who have not yet made their mark.

Robert K. Merton, "The Matthew Effect in Science,"
The Sociology of Science (1973)

Any fundamental theory of physics is beautiful. If it isn't, it's probably wrong.

John Moffat, in the University of Toronto *Bulletin,*
May 5, 1986

If I have seen further it is by standing on the shoulders of Giants.

> Sir Isaac Newton, in a letter to Robert Hooke,
> February 5, 1675 [See Gerald Holton and
> Francis Albert Eley Crew]

A new scientific truth does not triumph by convincing opponents and making them see the light, but rather because its opponents eventually die, and a new generation grows up that is familiar with it.

> Max Planck, *A Scientific Autobiography*
> *and Other Papers* (1949)

In art nothing worth doing can be done without genius; in science even a very moderate capacity can contribute to a supreme achievement.

> Bertrand Russell, "Science and Culture,"
> *Mysticism and Logic* (1917)

It would, of course, be a poor lookout for the advancement of science if young men started believing what their elders tell them, but perhaps it is legitimate to remark that young Turks look younger, or more Turkish . . . if the conclusions they eventually reach are different from what anyone had said before.

> Gunther Stent, in *Nature*, 1969

Beware of the problem of testing too many hypotheses;
the more you torture the data, the more likely they are to
confess, but confessions obtained under duress may not be
admissible in the court of scientific opinion.

> Stephen M. Stigler, "Testing Hypotheses or
> Fitting Models?" in *Neutral Models in Biology*
> edited by Matthew H. Nitecki and Antoni Hoffman (1987)

If you want to succeed in this world you don't have to be
much cleverer than other people, you just have to be one
day earlier.

> Leo Szilard, in *Leo Szilard: His Version of the Facts*
> edited by S. R. Weart and G. W. Szilard (1978)

The man with a new idea is a Crank until the idea succeeds.

> Mark Twain, "Pudd'nhead Wilson's New Calendar,"
> *Following the Equator* (1897)

The average scientist is good for at most one revolution.
Even if he has the power to make one change in his
category system and carry others along, success will make
him a recognized leader, with little to gain from another
revolution.

> Gerald M. Weinberg, *An Introduction to
> General Systems Thinking* (1975)

I was roused by the amphetamine of ambition.

Edward O. Wilson, on writing, *Sociobiology* (1975)

The scientist takes off from the manifold observations of predecessors, and shows his intelligence, if any, by his ability to discriminate between the important and the negligible, by selecting here and there the significant steppingstones that will lead across the difficulties to new understanding. The one who places the last stone and steps across to the terra firma of accomplished discovery gets all the credit.

Hans Zinsser, *As I Remember Him* (1940)

"

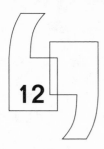

12

Scientists on Scientists

The world has arisen in some way or another. How it originated is the great question, and Darwin's theory, like all other attempts to explain the origin of life, is thus far merely conjectural. I believe he has not even made the best conjecture possible in the present state of our knowledge.

Louis Agassiz, *Evolution and Permanence of Type* (1874)

A biophysicist talks physics to the biologists and biology to the physicists, but when he meets another biophysicist, they just discuss women.

Anonymous, in *A Dictionary of Scientific Quotations*
by Alan L. Mackay (1991)

The astronomers must be very clever to have found out the names of all the stars.

Anonymous, in *The Physics Teacher*, October 1970

In the company of scientists, I feel like a shabby curate who has strayed . . . into a drawing room full of dukes.

W. H. Auden, "The Poet in the City"
The Dyer's Hand (1969)

. . . but Scientists who ought to know
Assure us [that] it must be so.
Oh, let us never, never doubt
What nobody is sure about.

Hilaire Belloc, "The Microbe" *TIBS,* July 1978

In many aspects, the theoretical physicist is merely a philosopher in a working suit.

P. Bergmann, in *Black Holes* by Jean-Pierre Luminet (1987)

No one believes an hypothesis except its originator but everyone believes an experiment except the experimenter. Most people are ready to believe something based on experiment but the experimenter knows the many little things that could have gone wrong in the experiment. For this reason the discoverer of a new fact seldom feels quite so confident of it as others do. On the other hand other people are usually critical of an hypothesis, whereas the originator identifies himself with it and is liable to become devoted to it.

W. I. B. Beveridge, *The Art of Scientific Investigation* (1950)

An expert is a man who has made all the mistakes, which can be made, in a very narrow field.

> Niels Bohr, recalled on his death, November 18, 1962

We all agreed that your theory is crazy. The question which divides us is whether it is crazy enough to have a chance of being correct.

> Niels Bohr, in *The Ambidextrous Universe*
> by Martin Gardner (1964)

Only those works which are well-written will pass to posterity: the amount of knowledge, the uniqueness of the facts, even the novelty of the discoveries are no guarantees of immortality. These things are exterior to a man but style is the man himself.

> Georges Leclarc Buffon, *Discours sur le Style* (1753)

The mere eminence of a specialist makes him the more dangerous.

> Alexis Carrel, *Man, the Unknown*

The best way of increasing the [average] intelligence of scientists would be to reduce their number.

> Alexis Carrel, *Man, the Unknown*

. . . the Einsteins were taken to the Mt. Wilson Observatory in California. Mrs. Einstein was particularly impressed by the giant telescope. "What on earth do they use it for?" she asked. Her host explained that one of its chief purposes was to find out the shape of the universe. "Oh," said Mrs. Einstein, "my husband does that on the back of an envelope."

Bennett Cerf, Try and Stop Me (1945)

When a distinguished but elderly scientist states that something is possible, he is almost certainly right. When he states that something is impossible, he is very probably wrong.

Arthur C. Clarke, in Time, February 15, 1971

Professors in every branch of the sciences prefer their own theories to truth: the reason is, that their theories are private property, but truth is common stock.

Charles Caleb Colton, Lacon (1849)

[Edison's ideas are] good enough for our transatlantic friends . . . but unworthy of the attention of practical or scientific men.

Committee set up by the British Parliament to look into
Edison's work on the incandescent lamp, c. 1878,
in Profiles of the Future by Arthur C. Clarke (1974)

"

There is only one proved method of assisting the advancement of pure science—that of picking men of genius, backing them heavily, and leaving them to direct themselves.

> James Bryant Conant, in a letter to
> *The New York Times,* August 13, 1945

When the views advanced by me in this volume . . . are generally admitted, we can dimly foresee that there will be a considerable revolution in natural history.

> Charles Darwin, *On the Origin of Species* (1859)

We don't do science for the general public. We do it for each other. Good day.

> Renato Dalbecco, the complete text of
> an interview with him in *The Sciences,* 1983

Most of the papers which are submitted to the *Physical Review* are rejected, not because it is impossible to understand them, but because it is possible. Those which are impossible to understand are usually published.

> Freeman J. Dyson, *Innovation in Physics*

[The National Academy of Sciences] would be unable to give a unanimous decision if asked whether the sun would rise tomorrow.

> Paul Ehrlich, in *Look,* April 1, 1970

In every true searcher of Nature there is a kind of religious reverence, for he finds it impossible to imagine that he is the first to have thought out the exceedingly delicate threads that connect his perceptions.

> Albert Einstein, 1920, in *Conversations with Einstein*
> by Alexander Moszkowski (1970)

If my theory of relativity is proven successful, Germany will claim me as a German and France will declare that I am a citizen of the world. Should my theory prove untrue, France will say that I am a German, and Germany will declare that I am a Jew.

> Albert Einstein, in a speech to the French Philosophical
> Society at the Sorbonne, April 6, 1922

It is an irony of fate that I myself have been the recipient of excessive admiration and reverence from my fellow-beings, through no fault and no merit of my own.

> Albert Einstein, "What I Believe," in *Forum and Century*, 1930

The years of searching in the dark for a truth that one feels but cannot express, the intense desire and the alternations of confidence and misgiving until one breaks through to clarity and understanding, are known only to him who has experienced them himself.

> Albert Einstein, in a lecture at the
> University of Glasgow, June 20, 1933

Why is that nobody understands me, and everybody likes me?

> Albert Einstein, in an interview in
> *The New York Times,* March 12, 1944

To punish me for my contempt for authority, Fate made me an authority myself.

> Albert Einstein, in *Albert Einstein: Creator and Rebel*
> by Banesh Hoffman (1972)

I sometimes ask myself how it came about that I was the one to develop the theory of relativity. The reason, I think, is that a normal adult never stops to think about problems of space and time. These are things which he has thought of as a child. But my intellectual development was retarded, as a result of which I began to wonder about space and time only when I had already grown up.

> Albert Einstein, in *Theories of Everything*
> by John D. Barrow (1991)

Before I came here I was confused about this subject. Having listened to your lecture I am still confused. But on a higher level.

> Enrico Fermi, in *A Dictionary of Scientific Quotations*
> by Alan L. Mackay (1991)

"

[Oppenheimer is] tense, dedicated, deeper than deep, somewhat haunted, uncertain, calm, confident, and full, full, full of knowledge, not only of particles and things but of men and motives, and of the basic humanity that may be the only savior we have in this strange world he and his colleagues have discovered.

> Philip Hamburger, in *Due to Circumstances Beyond Our Control* by Fred W. Friendly (1967)

[We must] recognize ourselves for what we are—the priests of a not very popular religion.

> Fred Hoyle, in *Physics Today,* April 1968

One machine can do the work of fifty ordinary men. No machine can do the work of one extraordinary man.

> Elbert Hubbard, *The Roycroft Dictionary and Book of Epigrams* (1923)

The improver of natural knowledge absolutely refuses to acknowledge authority, as such. For him, skepticism is the highest of duties, blind faith the one unpardonable sin.

> T. H. Huxley, *On the Advisableness of Improving Natural Knowledge* (1866)

I could more easily believe that two Yankee professors
would lie than that stones would fall from heaven.

> Thomas Jefferson, on a report on a meteorite shower
> which fell in Weston, Connecticut in 1807,
> in *Our Stone-pelted Planet* by H. H. Nininger (1933)

As marvelous as the stars is the mind of the person who
studies them.

> Martin Luther King, Jr., in *Voyage to the Great Attractor*
> by Alan Dressler (1995)

There were two kinds of physicists in Berlin: on the one
hand was Einstein, and on the other all the rest.

> Rudolph Ladenburg, in *Einstein: A Centenary Volume*
> edited by A. P. French (1979)

It's as important an event as would be the transfer of the
Vatican from Rome to the New World. The Pope of Physics
has moved and the United States will now become the
center of the natural sciences.

> Paul Langevin, on Albert Einstein's move to Princeton,
> New Jersey, from Germany in 1933,
> in *Brighter Than a Thousand Suns* by Robert Jungk (1958)

People give ear to an upstart astrologer [Copernicus] who
strove to show that the earth revolves, not the heavens or

the firmament, the sun and the moon. Whoever wishes to appear clever must devise some new system, which of all systems is of course the very best. This fool wishes to reverse the entire science of astronomy.

> Martin Luther, c. 1543, in *The Experts Speak*
> by Christopher Cerf and Victor Navasky (1998)

I can accept the theory of relativity as little as I can accept the existence of atoms and other such dogmas.

> Ernst Mach, Professor of Physics at the University of Vienna,
> 1913, in *The Book of Heroic Failures* by Stephen Pile (1979)

Inasmuch as science represents one way of dealing with the world, it does tend to separate its practitioners from the rest. Being a scientist resembles membership of a religious order and a scientist usually finds that he has more in common with a colleague on the other side of the world than with his next-door neighbor.

> Alan L. Mackay, *A Dictionary of Scientific Quotations* (1991)

Physicists are not regular fellows—and neither are poets. Anyone engaged in an activity that makes considerable demands on both the intellect and the emotions is not unlikely to be a little bit odd.

> Robert H. March, *Physics for Poets* (1978)

"

The value the world sets upon motives is often grossly unjust and inaccurate. Consider, for example, two of them: mere insatiable curiosity and the desire to do good. The latter is put high above the former, and yet it is the former that moves one of the most useful men the human race has yet produced: the scientific investigator. What actually urges him on is not some brummagem idea of Service, but a boundless, almost pathological thirst to penetrate the unknown, to uncover the secret, to find out what has not been found out before. His prototype is not the liberator releasing slaves, the good Samaritan lifting up the fallen, but a dog sniffing tremendously at an infinite series of rat-holes.

H. L. Mencken, in *Smart Set*, August 1919

Astronomers and physicists, dealing habitually with objects and quantities far beyond the reach of the senses, even with the aid of the most powerful aids that ingenuity has been able to devise, tend almost inevitably to fall into the ways of thinking of men dealing with objects and quantities that do not exist at all, e.g., theologians and metaphysicians.

H. L. Mencken, *Minority Report: H. L. Mencken's Notebook* (1956)

The wallpaper with which the men of science have covered the world of reality is falling to tatters.

Henry Miller, *The Tropic of Cancer* (1934)

"

It is rare to find learned men who are clean, do not stink and have a sense of humor.

> Charles-Louis de Secondat Montesquieu, in a letter to
> Sophie of Hanover about G. W. Leibnitz, July 30, 1705

If the whole of the English language could be condensed into one word, it would not suffice to express the utter contempt those invite who are so deluded as to be disciples of such an imposture as Darwinism.

> Francis Orpen Morris, British ornithologist,
> *Letters on Evolution* (1877)

How can he [Thomas Edison] call it a wonderful success when everyone acquainted with the subject will recognize it as a conspicuous failure?

> Henry Morton, Professor of Physics and President of the
> Stevens Institute of Technology, on Edison's incandescent
> light bulb, in *The New York Herald*, December 18, 1879

. . . all the great scientists have one thing in common: each snatched from the subtle motions of nature one irrevocable secret; each caught one feather of the plumage of the Great White Bird that symbolizes everlasting truth.

> Forest Ray Moulton and Justus J. Schifferes,
> *The Autobiography of Science* (1945)

I know not what I may appear to the world, but to myself I seem to have been only like a boy playing on the sea-shore, and diverting myself in now and then finding a smoother pebble or a prettier shell than ordinary, whilst the great ocean of truth lay all undiscovered before me.

> Sir Isaac Newton, in *Memoirs of the Life, Writings and Discoveries of Sir Isaac Newton,* Volume Two, by David Brewster (1860)

Both the man of science and the man of art live always at the edge of mystery, surrounded by it. Both, as the measure of their creation, have always had to do with the harmonization of what is new with what is familiar, with the balance between novelty and synthesis, with the struggle to make partial order out of chaos. . . . This cannot be an easy life.

> J. Robert Oppenheimer, in a lecture, 1954, in *Brighter than a Thousand Suns* by Robert Jungk (1958)

The true scientist never loses the faculty of amusement. It is the essence of his being.

> J. Robert Oppenheimer, in *The Peter Pyramid* by Laurence J. Peter (1986)

[Louis Pasteur's] . . . theory of germs is a ridiculous fiction. How do you think that these germs in the air can be

numerous enough to develop into all these organic infusions?
If that were true, they would be numerous enough to form
a thick fog, as dense as iron.

> Pierre Pochet, Professor of Physiology at Toulouse,
> *The Universe: The Infinitely Great and the*
> *Infinitely Small* (1872)

Nature, and Nature's laws lay hid in night;
God said, Let Newton be! And all was light.

> Alexander Pope, in *The Timeline Book of Science*
> by George Ochoa and Melinda Corey (1995)

Using any reasonable definition of a scientist, we can say
that 80 to 90 percent of all the scientists that have ever
lived are alive now. Alternatively, any young scientist,
starting now and looking back at the end of his career upon
a normal life span, will find that 80 to 90 percent of all
scientific work achieved by the end of the period will have
taken place before his very eyes, and that only 10 to 20
percent will antedate his experience.

> Derek J. de Solla Price, *Little Science, Big Science* (1963)

I think physicists are the Peter Pans of the human race.
They never grow up, and they keep their curiosity.

> I. I. Rabi, in *Experiencing Science* by Jeremy Bernstein (1978)

. . . the really fundamental things have a way of appearing
to be simple once they have been stated by a genius. . . .

> G. Y. Rainich, "Analytical Function and Mathematical Physics,"
> *Bulletin of the American Mathematical Society,* October 1931

Copernicus . . . did not publish his book [on the nature of
the solar system] until he was on his deathbed. He knew
how dangerous it is to be right when the rest of the world
is wrong.

> Thomas Brackett Reed, in a speech at
> Waterville, Maine, July 30, 1885

Not explaining science seems to me perverse. When you're
in love, you want to tell the world.

> Carl Sagan, in the *Washington Post,* January 9, 1994

I laughed . . . till my sides were sore.

> Adam Sedgwick, British geologist, describing,
> in a letter to Charles Darwin, his reaction to
> *On the Origin of Species,* in *Facts and Fallacies*
> by Chris Morgan and David Langford (1981)

It may well be that men of science, not kings, or warriors,
or even statesmen are to be the heroes of the future.

> William Robertson Smith and Charles A. Beard,
> *The Development of Modern Europe* (1908)

Scientists . . . I should say that naturally they had the future in their bones.

> C. P. Snow, *The Two Cultures and the Scientific Revolution* (1959)

Scientists have it within them to know what a future-directed society feels like, for science itself, in its human aspect, is just like that.

> C. P. Snow, *A Postscript to Science and Government* (1962)

The real scientist . . . is ready to bear privation and, if need be, starvation rather than let anyone dictate to him which direction his work must take.

> Albert Szent-Györgyi, "Science Needs Freedom," *World Digest,* 1943

It behooves us always to remember that in physics it has taken great men to discover simple things. They are very great names indeed which we couple with the explanation of the path of a stone, the droop of a chain, the tints of a bubble, the shadows of a cup.

> D'Arcy Wentworth Thompson, *On Growth and Form* (1917)

No doubt, a scientist isn't necessarily penalized for being a complex, versatile, eccentric individual with lots of extra-scientific interests. But it certainly doesn't help him a bit.

> Stephen Toulmin, *Civilization and Science in Conflict or Collaboration* (1972)

He [Leonardo da Vinci] might have been a scientist if he had not been so versatile.

> Giorgio Vasari, *Lives of the Artists* (1550)

. . . he could have gone around among his students with his penis hanging out, and everyone would have been charmed by his eccentricity.

> Johann von Neumann, on J. Robert Oppenheimer, in *The Fifties* by David Halberstam (1993)

Specialization has gotten out of hand. There are more branches in the tree of knowledge than there are in the tree of life. A petrologist studies rocks; a pedologist studies soils. The first one sieves the soil and throws away the rocks. The second one picks up the rocks and brushes off the soil. Out in the field, they bump into each other only like Laurel and Hardy, by accident, when they are both backing up.

> Jonathan Weiner, *The Next One Hundred Years* (1990)

"

Science has become adult; I am not sure whether scientists have.

> Victor Frederick Weisskopf, in *Scientists in Search of Their Conscience* edited by A. R. Michaelis and H. Harvey

Biologists can be just as sensitive to heresy as theologians.

> H. G. Wells, in *The International Dictionary of Thoughts*

A science which hesitates to forget its founders is lost.

> Alfred North Whitehead,
> in *A Dictionary of Scientific Quotations*
> by Alan L. Mackay (1991)

The independent scientist who is worth the slightest consideration as a scientist has a consecration which comes entirely from within himself: a vocation which demands the possibility of supreme self-sacrifice.

> Norbert Wiener, *The Human Use of Human Beings* (1950)

Cosmologists are often in error, but never in doubt.

> Ya.B. Zel'dovich, in *Light from the Depths of Time*
> by Rudolf Kippenhahn (1987)

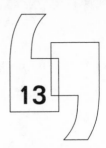

13

Humankind's Place in the Universe

[The great supercomputer, asked what is the answer to] the great problems of life, the universe and everything [replied, after many years of computation] 42.

Douglas Adams, *The Hitchhiker's Guide to the Galaxy* (1979)

There is a theory which states that if ever anyone discovers exactly what the Universe is for and why it is here, it will instantly disappear and be replaced by something even more bizarre and inexplicable.

There is another which states that this has already happened.

Douglas Adams, *The Restaurant at the End of the Universe* (1980)

Mother: He's been depressed. All of a sudden, he can't do
          anything.
Doctor: Why are you depressed, Alvy?
Mother: Tell Dr. Flicker. It's something he read.
Doctor: Something he read, huh?
Alvy: The universe is expanding.
Doctor: The universe is expanding?
Alvy: Well, the universe is everything, and if it's
          expanding, someday it will break apart and
          that would be the end of everything!
Mother: What is that your business? He stopped doing his
          homework.
Alvy: What's the point?

> Woody Allen and Marshall Brickman, *Annie Hall* (1973)

That's one small step for a man, one giant leap for
mankind.

> Neil Armstrong, on becoming the first person to stand
> on the moon, July 20, 1969 [See Lily Tomlin]

As far as the meaning of life in general, or in the abstract,
as far as I can see, there is none. If all of life were suddenly
to disappear from earth and anywhere else it may exist, or
if none had ever formed in the first place, I think the Universe would continue to exist without perceptible change.
However, it is always possible for an individual to invest his
own life with meaning that he can find significant. He can
so order his life that he may find as much beauty and

wisdom in it as he can, and spread as much of that to others as possible.

Isaac Asimov, in a book proposal for *The Meaning of Life* edited by Hugh S. Moorhead, 1989

Nature, to be commanded, must be obeyed.

Sir Francis Bacon, *Novum Organum* (1620)

There is no reason that the universe should be designed for our convenience.

John D. Barrow, *The Origin of the Universe* (1994)

Man is occupied and has been persistently occupied since his separate evolution, with three kinds of struggle: first with the massive unintelligent forces of nature, heat and cold, winds, rivers, matter and energy; secondly, with the things closer to him, animals and plants, his own body, its health and disease; and lastly, with his desires and fears, his imaginations and stupidities.

John Desmond Bernal,
The World, the Flesh and the Devil (1929)

Life is a partial, continuous, progressive, multiform and conditionally interactive, self-realization of the potentialities of atomic electron states.

John Desmond Bernal, *The Origin of Life* (1967)

Since nothing stands in the way of the movability of the earth, I believe we must now investigate whether it also has several motions, so that it can be considered one of the planets.

Nicolaus Copernicus,
De Revolutionibus Orbium Coelestium (1543)

A man said to the universe:
"Sir, I exist!"
"However," replied the universe,
"The fact has not created in me
A sense of obligation."

Stephen Crane, "A Man Said to the Universe,"
War Is Kind (1899)

You and I are flesh and blood, but we are also stardust.

Helena Curtis, on the formation of the planet earth,
Biology (1968)

As we look out into the Universe and identify the many accidents of physics and astronomy that have worked together to our benefit, it almost seems as if the Universe must in some sense have known that we were coming.

Freeman J. Dyson, in *The Anthropic Cosmological Principle*
by John D. Barrow and Frank J. Tipler (1986)

We have found a strange footprint on the shores of
the unknown. We have devised profound theories, one
after another, to account for its origin. At last, we have
succeeded in reconstructing the creature that made the
footprint. And lo! It is our own.

Sir Arthur Stanley Eddington,
Space, Time and Gravitation (1959)

I will give you a "celestial multiplication table." We start
with a star as the unit most familiar to us, a globe
comparable to the sun.
Then—
A hundred thousand million Stars make one Galaxy;
A hundred thousand million Galaxies make one Universe.

Sir Arthur Stanley Eddington,
The Expanding Universe (1933)

What is the meaning of human life, or for that matter,
of the life of any creature? To know an answer to this
question means to be religious. Does it make any sense,
then, to pose this question? I answer: The man who regards
his own life and that of his fellow creatures as meaningless
is not merely unhappy but hardly fit for life.

Albert Einstein, *Ideas and Opinions* (1954)

The universe does not jest with us, but is in earnest.

Ralph Waldo Emerson, Journal, 1841

. . . I should think that anyone who considered it more reasonable for the whole universe to move in order to let the Earth remain fixed would be more irrational than one who should climb to the top of your cupola just to get a view of the city and its environs, and then demand that the whole countryside should revolve around him so that he would not have to take the trouble to turn his head.

Galileo Galilei, *Dialogue Concerning the Two Chief World Systems* (1632)

Man is not born to solve the problems of the universe, but to find where the problems begin, and then to take his stand within the limits of the intelligible.

Johann Wolfgang von Goethe, in *Exploring the Cosmos* by Louis Berman (1973)

. . . if we do discover a complete [unified] theory [of the universe], it should in time be understandable in broad principle by everyone, not just a few scientists. Then we shall all, philosophers, scientists, and just ordinary people, be able to take part in the discussion of the question of why it is that we and the universe exist. If we find the answer to that, it would be the ultimate triumph of human reason— for then we would know the mind of God.

Stephen Hawking, *A Brief History of Time* (1988)

We see the universe the way it is because if it were different, we would not be here to observe it.

Stephen Hawking, in the *Washington Post,* April 15, 1988

The universe is not hostile, nor yet is it friendly. It is simply indifferent.

John Haynes Holmes,
A Sensible Man's View of Religion (1932)

The chess-board is the world; the pieces are the phenomena of the universe; the rules of the game are what we call the laws of Nature. The player on the other side is hidden from us. We know that his play is always fair, just, and patient. But also we know, to our cost, that he never overlooks a mistake, or makes the smallest allowance for ignorance.

T. H. Huxley, "A Liberal Education,"
Lay Sermons, Addresses, and Reviews (1871)

Myths and science fulfill a similar function: they both provide human beings with a representation of the world and of the forces that are supposed to govern it. They both fix the limits of what is considered as possible.

François Jacob, *The Possible and the Actual* (1982)

Astronomy is perhaps the science whose discoveries owe least to chance, in which human understanding appears in its whole magnitude, and through which man can best learn how small he is.

Georg Christoph Lichtenberg, Lichtenberg:
Aphorisms and Letters (1969)

Although we are mere sojourners on the surface of the planet, chained to a mere point in space, enduring but for a moment of time, the human mind is not only enabled to number worlds beyond the unassisted ken of mortal eye, but to trace the events of indefinite ages before the creation of our race, and is not even withheld from penetrating into the dark secrets of the ocean, or the interior of the solid globe; free, like the spirit which the poet described as animating the universe.

Sir Charles Lyell, Principles of Geology (1830)

Biology occupies a position among the sciences at once marginal and central. Marginal because—the living world constituting but a tiny and very "special" part of the universe—it does not seem likely that the study of living beings will ever uncover general laws applicable outside the biosphere. But if the ultimate aim of the whole of science is indeed, as I believe, to clarify man's relationship to the universe, then biology must be accorded a central position. . . .

Jacques Monod, Chance and Necessity (1971)

There is, then, one great purpose for man and for us today, and that is to try to discover man's purpose by every means in our power. That is the ultimate relevance of science, and not only of science, but of every branch of learning which can improve our understanding. In the words of Tolstoy, "The highest wisdom has but one science, the science of the whole, the science explaining the Creation and man's place in it."

Sir George Porter, *The Times,* June 21, 1975

The universe forces those who live in it to understand it.

Carl Sagan, *Broca's Brain: Reflections on the Romance of Science* (1979)

The size and age of the Cosmos are beyond ordinary human understanding. Lost somewhere between immensity and eternity is our tiny planetary home.

Carl Sagan, *Cosmos* (1980)

I would rather believe that God did not exist than believe that He was indifferent.

George Sand (Armadine Aurore Lucile Dupin), *Impressions et Souvenirs* (1896)

Within our bodies course the same elements that flame in the stars.

Susan Schiefelbein, *The Incredible Machine* (1986)

"

But I canna change the laws of physics, Captain!

> Montgomery Scott ("Scotty"), chief engineer aboard the
> U.S.S. *Enterprise,* to Captain James T. Kirk on *Star Trek*

Nature holds no brief for the human experiment; it must
stand or fall by its results.

> George Bernard Shaw, Preface to
> *Back to Methuselah* (1921)

We are all but recent leaves on the same old tree of life
and if this life has adapted itself to new functions and
conditions, it uses the same old basic principles over and
over again. There is no real difference between the grass
and the man who mows it.

> Albert Szent-Györgyi, in *Free Radical* by R.W. Moss (1988)

I do not value any view of the universe into which man and
the institutions of man enter very largely and absorb much
of the attention. Man is but the place where I stand, and
the prospect hence is infinite.

> Henry David Thoreau, Journal, April 2, 1852

The astronauts go to the moon, and what do they do?
They collect rocks, they lope around like they are stoned,
and they hit a golf ball and they plant a flag. . . . All that
technology to get to the moon, and what do we do—we
play golf. . . . The moon walk was out of sight, wasn't it?

With one giant step mankind took banality out of America
and into the Cosmos. . . .

Lily Tomlin, "Appearing Nightly," 1980 [See Neil Armstrong]

And beyond our galaxy are other galaxies, in the universe
all told at least a hundred billion, each containing a hundred
billion stars. Do these figures mean anything to you?

John Updike, *The Centaur* (1990)

In man's brain the impressions from outside are not merely
registered; they produce concepts and ideas. They are the
imprint of the external world upon the human brain.
Therefore, it is not surprising that, after a long period
of searching and erring, some of the concepts and ideas in
human thinking should have come gradually closer to the
fundamental laws of the world, that some of our thinking
should reveal the true structure of atoms and the true
movements of the stars. Nature, in the form of man,
begins to recognize itself.

Victor Frederick Weisskopf, *Knowledge and Wonder* (1962)

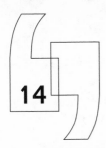

14

. . . and All Else Under the Heavens

In the beginning the universe was created. This made a lot of people very angry and [has] been widely regarded as a bad move. Many races believe it was created by some sort of god, though the Jatravartid people of Viltvodle VI believe that the entire Universe was in fact sneezed out of the nose of a being called the Great Green Arkleseizure.

> Douglas Adams, *The Restaurant at the End of the Universe* (1980)

If the Lord Almighty had consulted me before embarking upon creation, I should have recommended something simpler.

> Alfonso X (King of Castile and Leon, 1252–84), on having the Ptolemaic system of astronomy explained to him, in *A Dictionary of Scientific Quotations* by Alan L. Mackay (1991)

If You Build Your House on a Crack in the Earth, It's Your Own Fault.

> Anonymous, title of a scientific paper on earthquakes, in *Peter's People* by Laurence J. Peter (1979)

If . . . the motion of the earth were circular, it would be violent and contrary to nature, and could not be eternal since . . . nothing violent is eternal. . . . It follows, therefore, that the earth is not moved with a circular motion.

> Saint Thomas Aquinas, *Commentaria in Libros Aristotelis de Caelo et Mundo* (c. 1270)

If one way be better than another, that you may be sure is Nature's way.

> Aristotle, *Nichomachean Ethics* (4th century B.C.)

Houston, Tranquillity Base here. The *Eagle* has landed.

> Neil Armstrong, first words from the moon, July 20, 1969

If God meant for us to fly he would have given us propellers on our noses.

> Isaac Asimov, on *The Dick Cavett Show,* February 24, 1971

Remember that accumulated knowledge, like accumulated capital, increases at compound interest: but it differs from the accumulation of capital in this; that the increase of knowledge produces a more rapid rate of progress, whilst the accumulation of capital leads to a lower rate of interest. Capital thus checks its own accumulation: knowledge thus accelerates its own advance. Each generation, therefore, to deserve comparison with its predecessor, is bound to add much more largely to the common stock than that which it immediately succeeds.

Charles Babbage, *The Exposition of 1851* (1851)

He that will not apply new remedies must expect new evils; for time is the greatest innovator.

Sir Francis Bacon, "On Innovation," *Essays* (1597)

It is possible that some other science may be more useful [than optics], but no other science has so much sweetness and beauty of utility. Therefore it is the flower of the whole of philosophy and through it, and not without it, can the other sciences be known.

Roger Bacon, *Opus Majus* (1267)

In a randomly infinite Universe, any event occurring here and now with finite probability must be occurring simultaneously at an infinite number of other sites in the Universe. It is hard to evaluate this idea any further, but one thing is certain: if it is true then it is certainly not original!

John Barrow and Frank Tipler,
The Anthropic Cosmological Principle (1986)

"

What is found in the effect was already in the cause.

> Henri Bergson, *Creative Evolution* (1907)

No longer can we be satisfied with a life where the heart has its reasons which reason cannot know. Our hearts must know the world of reason, and reason must be guided by an informed heart.

> Bruno Bettelheim, in *The Guardian*, March 15, 1990

Energy is the only life . . . as Reason is the bound or outward circumference of Energy.

> William Blake, *The Marriage of Heaven and Hell* (1792–93) [See Freeman J. Dyson]

When it comes to atoms, language can be used only as in poetry. The poet too is not nearly so concerned with describing facts as with creating images.

> Niels Bohr, in *The Ascent of Man* by Jacob Bronowski (1975)

Anyone who is not shocked by quantum theory has not understood it.

> Niels Bohr, in *Maya in Physics* by N. C. Panda (1991)

DNA was the first three-dimensional Xerox machine.

> Kenneth Boulding, "Energy and the Environment," *Beasts, Ballads, and Bouldingisms* (1976)

I am selling what the whole world wants: power.

> Matthew Boulton, in a letter to Catherine the Great of Russia offering steam engines for sale, in *A Dictionary of Scientific Quotations* by Alan L. Mackay (1991)

The idea that the universe is running down comes from a simple observation about machines. Every machine consumes more energy than it renders.

> Jacob Bronowski, *The Ascent of Man* (1973)

To treat your facts with imagination is one thing, to imagine your facts is quite another.

> John Burroughs, Journal, October 24, 1907

I am tired of all this thing called science. . . . We have spent millions on that sort of thing for the last few years, and it is time it should be stopped.

> Senator Simon Cameron, demanding that the funding of the Smithsonian Institution be cut off, 1861, in *Random Walk in Science* by R. L. Weber and E. Mendoza (1973)

Often a liberal antidote of experience supplies a sovereign cure for a paralyzing abstraction built upon a theory.

> Benjamin N. Cardozo, *The Paradoxes of Legal Science* (1928)

Tyrannosaurus was truly the Schwarzenegger of dinosaurs.

> Kenneth Carpenter, in *The New York Times,* July 3, 1990

. . . to the bourgeois . . . art and science appear not as creative opposites but as eternal antagonists.

Christopher Caudwell, *Studies in a Dying Culture* (1938)

The works of Nature must all be accounted good.

Cicero, *De Senectute* (1st century B.C.)

All revolutionary advances in science may consist less of sudden and dramatic revelations than a series of transformations, of which the revolutionary significance may not be seen (except afterwards, by historians) until the last great step. In many cases the full potentiality and force of a most radical step in such a sequence of transformations may not even be manifest to its author.

I. Bernard Cohen, *The Newtonian Revolution* (1980)

Wooden legs are not inherited but wooden heads may be.

Edwin G. Conkling, on the view that acquired characteristics can be inherited, recalled on his death, November 21, 1952

Chance is the only source of true novelty.

Francis Harry Compton Crick, *Life Itself* (1982)

To suppose that the eye with all its inimitable contrivances for adjusting the focus to different distances, for admitting different amounts of light, and for the correction of spherical and chromatic aberration, could have been formed by natural selection, seems, I confess, absurd in the highest degree.

Charles Darwin, On the Origin of Species (1859)

The preservation of favorable variations and the rejection of injurious variations, I call Natural Selection, or Survival of the Fittest. Variations neither useful nor injurious would not be affected by natural selection and would be left a fluctuating element.

Charles Darwin, On the Origin of Species (1859)

Although gravity is by far the weakest force of nature, its insidious and cumulative action serves to determine the ultimate fate not only of individual astronomical objects but of the entire cosmos. The same remorseless attraction that crushes a star operates on a much grander scale on the universe as a whole.

Paul Davies, The Last Three Minutes (1994)

Nothing tends so much to the advancement of knowledge as the application of a new instrument. The native intellectual powers of men in different times are not so much the causes of the different success of their labors, as the peculiar nature of the means and artificial resources in their possession.

Sir Humphrey Davy, in Force of Nature
by Thomas Hager (1995)

"

How can we be so willfully blind as to look for causes in nature when nature herself is an effect?

Joseph-Marie de Maistre

I think, therefore I am.

René Descartes, *Discourse on Method* (1637)

Putting on the spectacles of science in expectation of finding the answer to everything looked at signifies inner blindness.

J. Frank Dobie, *The Voice of the Coyote* (1949)

Nothing in Biology Makes Sense Except in the Light of Evolution.

Theodosius Gregorievich Dobzhansky, title of article in *American Biology Teacher*, 1973

From a drop of water a logician could predict an Atlantic or a Niagara.

Sir Arthur Conan Doyle, *A Study in Scarlet* (1887)

We do not know how the scientists of the next century will define energy or in what strange jargon they will discuss it. But no matter what language the physicists use they will not come into contradiction with Blake. Energy will remain in some sense the lord and giver of life, a reality transcending our mathematical descriptions. Its nature lies at the heart of

the mystery of our existence as animate beings in an
inanimate universe.

> Freeman J. Dyson, "Energy in the Universe,"
> *Scientific American,* September 1971 [See William Blake]

He that increaseth knowledge, increaseth sorrow.

> Ecclesiastes 1:18

A happy man is too satisfied with the present to dwell too
much on the future.

> Albert Einstein, "My Future Plans," September 18, 1896

He who has never been deceived by a lie does not know the
meaning of bliss.

> Albert Einstein, in a letter to Elsa Löwenthal, April 30, 1912

It is not so very important for a person to learn facts. For
that he does not really need a college. He can learn them
from books. The value of an education in a liberal arts
college is not the learning of many facts but the training of
the mind to think something that cannot be learned from
textbooks.

> Albert Einstein, 1921, commenting on
> Thomas Edison's opinion that a college education is useless,
> in *Einstein: His Life and Times* by Philipp Frank (1953)

Politics is a pendulum whose swings between anarchy and tyranny are fueled by perennially rejuvenated illusions.

> Albert Einstein, 1937, in *Albert Einstein, the Human Side*
> by Helen Dukas and Banesh Hoffman (1979)

To act intelligently in human affairs is only possible if an attempt is made to understand the thoughts, motives, and apprehensions of one's opponent so fully that one can see the world through his eyes.

> Albert Einstein, in *The New York Times*, 1948

When I study philosophical works I feel I am swallowing something which I don't have in my mouth.

> Albert Einstein, in *A Dictionary of Scientific Quotations*
> by Alan L. Mackay (1991)

Every chemical substance, every plant, every animal in its growth, teaches the unity of the cause, the variety of appearance.

> Ralph Waldo Emerson, "History," *Essays: First Series* (1841)

Knowledge is a process of piling up facts; wisdom lies in their simplification.

> Harold Fabing and Ray Marr, *Fischerisms* (1937)

Think, for a moment, of a cheetah, a sleek, beautiful animal, one of the fastest on earth, which roams freely on the savannas of Africa. In its natural habitat, it is a magnificent animal, almost a work of art, unsurpassed in speed or grace by any other animal. Now, think of a cheetah that has been captured and thrown into a miserable cage in a zoo. It has lost its original grace and beauty, and is put on display for our amusement. We see only the broken spirit of the cheetah in the cage, not its original power and elegance. The cheetah can be compared to the laws of physics, which are beautiful in their natural setting. The natural habitat of the laws of physics is a higher-dimensional space-time. However, we can only measure the laws of physics when they have been broken and placed on display in a cage, which is our three-dimensional laboratory. We only see the cheetah when its grace and beauty have been stripped away.

> Peter Freund, in *Hyperspace* by Michio Kaku (1994)

There are three great things in the world: there is religion, there is science, and there is gossip.

> Robert Frost, 1963, in *The Timeline Book of Science*
> by George Ochoa and Melinda Corey (1995)

It is a commonplace of modern technology that problems have solutions before there is knowledge of how they are to be solved.

> John Kenneth Galbraith, *The New Industrial State* (1977)

"

Anyone who does not grasp the close juxtaposition of the vulgar and the scholarly has either too refined or too compartmentalized a view of life. Abstract and visceral fascination are equally valid and not so far apart.

Stephen Jay Gould, *The Flamingo's Smile* (1985)

The conservative has but little to fear from the man whose reason is the servant of his passions, but let him beware of him in whom reason has become the greatest and most terrible of the passions.

J. B. S. Haldane, *Daedalus, or Science and the Future* (1923)

The universe consists only of atoms and the void: all else is opinion and illusion.

Edward Robert Harrison, *Masks of the Universe* (1985)

We need not hesitate to admit that the Sun is richly stored with inhabitants.

Sir William Herschel, Court Astronomer of England and discoverer of the planet Uranus, 1781, in *The Comet Is Coming* by Niegel Calder (1980)

Experimental evidence is strongly in favor of my argument that the chemical purity of the air is of no importance.

L. Erskine Hill, Lecturer on Physiology at London Hospital, in "Impure Air Not Unhealthful If Stirred and Cooled," in *The New York Times*, September 22, 1912

There are only two things, science and opinion; the former begets knowledge, the latter ignorance.

> Hippocrates, *Law* (5th–4th century B.C.)

As children we all possess a natural, uninhibited curiosity, a hunger for explanation, which seems to die slowly as we age—suppressed, I suppose, by the high value we place on conformity and by the need not to appear ignorant. It betokens a conviction that somehow science is innately incomprehensible. It precludes reaching deeper, thereby denying the profound truth that understanding enriches experience, that explanation vastly enhances the beauty of the natural world in the eye of the beholder.

> Mahlon Hoagland, *Toward the Habit of Truth* (1990)

Nature itself cannot err.

> Thomas Hobbes, *Leviathan* (1651)

It is the province of knowledge to speak and it is the privilege of wisdom to listen.

> Oliver Wendell Holmes, Sr.,
> *The Poet at the Breakfast-Table* (1892)

Facts are ventriloquist's dummies. Sitting on a wise man's knee they may be made to utter words of wisdom; elsewhere they say nothing, or talk nonsense.

> Aldous Huxley, *Time Must Have a Stop* (1945)

"

. . . if a little knowledge is a dangerous thing, where is the man who has so much as to be out of danger?

> T. H. Huxley, "On Elementary Instruction in Physiology,"
> *Collected Essays* (1903)

The Universe was a stage in which always the same actors—the atoms—played their parts, differing in disguises and groupings, but without change of identity. And these actors were endowed with immortality.

> Sir James Jeans, *The Mysterious Universe* (1948)

The greater the tension, the greater the potential. Great energy springs from a correspondingly great tension of opposites.

> Carl G. Jung, "Paracelsus as a Spiritual Phenomenon,"
> *Alchemical Studies* (1967)

In order to drive the individuals towards reproduction, sexuality had therefore to be associated with some other devices. Among these was pleasure. . . . Thus pleasure appears as a mere expedient to push individuals to indulge in sex and therefore to reproduce. A rather successful expedient indeed as judged by the state of the world population.

> François Jacob, "Evolution and Tinkering,"
> *Science,* June 10, 1977

It is often stated that of all the theories proposed in this century, the silliest is quantum theory. In fact, some say that the only thing that quantum theory has going for it is that it is unquestionably correct.

Michio Kaku, *Hyperspace* (1995)

Two things fill the mind with ever new and increasing admiration and awe, the oftener and more steadily they are reflected on: the starry heavens above me, and the moral law within me.

Immanuel Kant, *Critique of Practical Reason* (1788)

So long as the mother, Ignorance, lives, it is not safe for Science, the offspring, to divulge the hidden causes of things.

Johannes Kepler, *Somnium* (1634)

Mother Astronomy would surely have to suffer hunger if the daughter Astrology did not earn their bread.

Johannes Kepler, on his having to cast horoscopes to make a living, in *The Autobiography of Science* edited by Forest Ray Moulton and Justus J. Schifferes (Second Edition) (1960)

The difficulty lies, not in the new ideas, but in escaping the old ones, which ramify, for those brought up as most of us have been, into every corner of our minds.

John Maynard Keynes, in *Engines of Creation* by K. Eric Drexler (1987)

"

The understanding of atomic physics is child's play compared with the understanding of child's play.

> David Kresch, in *A Dictionary of Scientific Quotations*
> by Alan L. Mackay (1991)

Though many have tried, no one has ever yet explained away the decisive fact that science, which can do so much, cannot decide what it ought to do.

> Joseph Wood Krutch, "The Loss of Confidence,"
> *The Measure of Man* (1954)

It is impossible to disassociate language from science or science from language, because every natural science always involves three things: the sequence of phenomena on which the science is based; the abstract concepts which call these phenomena to mind; and the words in which the concepts are expressed. To call forth a concept a word is needed; to portray a phenomenon a concept is needed. All three mirror one and the same reality.

> Antoine Lavoisier, *Traité Elémentaire de Chimie* (1789)

We hope to explain the universe in a single, simple formula that you can wear on your T-shirt.

> Leon Lederman, in "Quark City"
> by Richard Wolkomir, *Omni,* February 1984

Nature does not make jumps.

Carl Linnaeus, *Philosophia Botanica* (1751)

Significant advances in science often have a peculiar quality: they contradict obvious, commonsense opinions.

S. E. Luria, *A Slot Machine, A Broken Test Tube:
An Autobiography* (1984)

Where do correct ideas come from? Do they drop from the skies? No. Are they innate in the mind? No. They come from social practice, and from it alone; they come from three kinds of social practice, the struggle for production, the class struggle and scientific experiment.

Mao Tse-Tung, "Where Do Correct Ideas Come From?"
Quotations from Chairman Mao Tse-Tung (1967)

If you see a formula in the *Physical Review* that extends over a quarter of a page, forget it. It's wrong. Nature isn't that complicated.

Bernd T. Matthias, Professor of Physics, University of
California at San Diego, in *A Dictionary of
Scientific Quotations* by Alan L. Mackay (1991)

People who write obscurely are either unskilled in writing or up to mischief.

Sir Peter Medawar, *Science and Literature in
Plato's Republic* (1984)

"

However far modern science and technics have fallen short of their inherent possibilities, they have taught mankind at least one lesson: Nothing is impossible.

> Lewis Mumford, *Technics and Civilization* (1934)

What was once called the objective world is a sort of Rorschach ink blot, into which each culture, each system of science and religion, each type of personality, reads a meaning only remotely derived from the shape and color of the blot itself.

> Lewis Mumford, "Orientation to Life," *The Conduct of Life* (1951)

Our exploration of the planets represents a triumph of imagination and will for the human race. The events of the last twenty years are perhaps too recent for us to adequately appreciate their proper historical significance. We can, however, appraise the scientific significance of these voyages of exploration: They have been nothing less than revolutionary both in providing a new picture of the nature of the solar system, its likely origin and evolution, and in giving us a new perspective on our own planet Earth.

> NASA Advisory Committee, report of Solar System Exploration Committee, *Planetary Exploration Through Year 2000: A Core Program* (1983)

Laboratorium est oratorium. [The place where we do our scientific work is a place of prayer.]

Joseph Needham, in *A Dictionary of Scientific Quotations*
by Alan L. Mackay (1991)

If each of us can be helped by science to live a hundred years, what will it profit us if our hates and fears, our loneliness and our remorse will not permit us to enjoy them? What use is an extra year or two to the man who "kills" what time he has?

David Neiswanger

That one body may act on another through a vacuum, without the mediation of anything else, by and through which their action and force may be conveyed from one to another, is to be so great an absurdity, that I believe no man who has in philosophical matters a competent faculty of thinking can ever fall into it.

Sir Isaac Newton, in a letter to Richard Bentley, c. 1692,
in *A History of Astronomy* by A. Pannekoek (1961)

I'm not afraid of facts, I welcome facts *but a congeries of fact is not equivalent to an idea.* This is the essential fallacy of the so-called "scientific" mind. People who mistake facts for ideas are incomplete thinkers; they are gossips.

Cynthia Ozick, "We Are the Crazy Lady and Other Feisty
Feminist Fables," in *The First Ms. Reader*
edited by Francine Klagsburn (1972)

"

In the fields of observation chance favors only the prepared mind.

> Louis Pasteur, in a lecture at the University of Lille, December 7, 1854

Freedom is for science what the air is for an animal.

> Henri Poincaré, *Dernières Pensées*

Nature composes some of her loveliest poems for the microscope and the telescope.

> Theodore Roszak, *Where the Wasteland Ends* (1972)

Even if the open windows of science at first make us shiver after the cozy indoor warmth of traditional humanizing myths, in the end the fresh air brings vigor, and the great spaces have a splendor of their own.

> Bertrand Russell, *What I Believe* (1925)

A machine is not a genie, it does not work by magic, it does not possess a will, and . . . nothing comes out which has not been put in, barring of course, an infrequent case of malfunctioning. . . . The "intentions" which the machine seems to manifest are the intentions of the human programmer, as specified in advance, or they are subsidiary intentions derived from these, following rules specified by the programmer. . . . The machine will not and cannot do any of these things until it has been instructed as to how to proceed. . . . To believe otherwise is either to believe in magic or to believe that the existence of man's will is an illusion and that man's actions are as mechanical as the machine's.

Arthur L. Samuel, in Science, September 16, 1960

The whole secret of the study of nature lies in learning how to use one's eyes.

George Sand (Armadine Aurore Lucile Dupin),
Nouvelles Lettres d'un Voyageur (1869)

All problems are finally scientific problems.

George Bernard Shaw, Preface, The Doctor's Dilemma (1911)

The real problem is not whether machines think but whether men do.

B. F. Skinner, Contingencies of Reinforcement (1969)

Be enthusiastic. Remember the placebo effect—30% of medicine is showbiz.

Ronald Spark, advising his medical colleagues,
in Medical World News, February 16, 1981

It is the nature of a hypothesis, when once a man has conceived it, that it assimilates every thing to itself, as proper nourishment; and, from the first moment of your begetting it, generally grows the stronger by every thing you see, hear, read, or understand.

> Laurence Sterne, *The Life and Opinions of Tristram Shandy* (1760)

Knowledge is a sacred cow, and my problem will be how we can milk her while keeping clear of her horns.

> Albert Szent-Györgyi, "Teaching and Expanding Knowledge," *Science,* December 4, 1964

The history of the living world can be summarized as the elaboration of ever more perfect eyes within a cosmos in which there is always something more to be seen.

> Pierre Teilhard de Chardin, *The Phenomenon of Man* (1955)

Scientific wealth tends to accumulate according to the law of compound interest. Every addition to knowledge of the properties of matter supplies the [physical scientist] with new instrumental means of discovering and interpreting phenomena of nature, which in their turn afford foundations of fresh generalizations, bringing gains of permanent value into the great storehouse of [natural] philosophy.

> William Thomson (Lord Kelvin), in a speech to the British Association, 1871

There are three kinds of lies—lies, damned lies and statistics.

> Mark Twain, *Autobiography* (1924)

The researches of many commentators have already thrown much darkness on this subject, and it is probable that, if they continue, we shall soon know nothing at all about it.

> Mark Twain, in *The Sciences,* September–October 1989

The brightest flashes in the world of thought are incomplete until they have been proved to have their counterparts in the world of fact.

> John Tyndall, "Scientific Materialism,"
> *Fragments of Science for Unscientific People* (1871)

Happy is he who has been able to learn the causes of things.

> Virgil, *Georgics* (36–29 B.C.)

It would be a poor thing to be an atom in a universe without physicists, and physicists are made of atoms. A physicist is an atom's way of knowing about atoms.

> George Wald, in *The Fitness of the Environment*
> by L. J. Henderson (1958)

It is a test of true theories not only to account for but to predict phenomena.

> William Whewell, *Philosophy of the*
> *Inductive Sciences* (1840)

Index of Authors